Gözde Gözke

Electrofiltration of Biopolymers

Spatially Distributed Process Analysis

Electrofiltration of Biopolymers

Spatially Distributed Process Analysis

by
Gözde Gözke

Dissertation, Karlsruher Institut für Technologie
Fakultät für Chemieingenierwesen und Verfahrenstechnik
Tag der mündlichen Prüfung: 30.03.2012
Referenten: Prof. Dr.-Ing. Clemens Posten; Prof. Dr. Ursula Obst

Impressum

Karlsruher Institut für Technologie (KIT)
KIT Scientific Publishing
Straße am Forum 2
D-76131 Karlsruhe
www.ksp.kit.edu

KIT – Universität des Landes Baden-Württemberg und nationales
Forschungszentrum in der Helmholtz-Gemeinschaft

KIT Scientific Publishing 2012
Print on Demand

ISBN 978-3-86644-845-2

Electrofiltration of biopolymers

Spatially distributed process analysis

zur Erlangung des akademischen Grades eines
DOKTORS DER INGENIEURWISSENSCHAFTEN (Dr.-Ing.)

der Fakultät für Chemieingenieurwesen und Verfahrenstechnik des
Karlsruher Instituts für Technologie (KIT)

genehmigte
DISSERTATION

von
M.Sc. Gözde Gözke
aus Istanbul

Referent: Prof. Dr.-Ing. Clemens Posten
Korreferentin: Prof. Dr. Ursula Obst
Tag der mündlichen Prüfung: 30.03.2012

Danksagung

Die vorliegende Dissertation entstand während meiner Tätigkeit als wissenschaftliche Mitarbeiterin am Institut für Bio- und Lebensmitteltechnik Abteilung Bioverfahrenstechnik des Karlsruher Instituts für Technologie (KIT) in der Zeit von 2006 bis 2011.

Mein besonderer Dank gilt Ihnen Herr Prof. Dr.-Ing. Clemens Posten für die wissenschaftliche Betreuung dieser Arbeit. Ihr gewährtes Vertrauen und unsere fruchtbaren Diskussionen trugen sehr zum Gelingen meiner Arbeit bei und erweiterten meinen Horizont.

Frau Prof. Dr. Ursula Obst danke ich herzlich für die freundliche Übernahme des Korreferats.

Meinen Kollegen möchte ich für die gute Zusammenarbeit und die nette Arbeitsatmosphäre danken. Insbesondere danke ich den Kollegen in der Partikelgruppe Herrn Dr.-Ing. Nikolay Krumov für die Hilfe beim Editieren der Publikationen, Frau Dr. Meike Dössel für die fachliche Unterstützung in chemischen Fragestellungen, Frau Dr.-Ing. Iris Perner-Nochta für die kompetenten Anregungen, Herrn Dipl.-Ing. Alexander Scholz und Herrn Dipl.-Ing. Martin Cerff für ihre stete Hilfsbereitschaft. Bei Herrn Dipl.-Ing. Christian Steinweg und Frau B.Sc. Yvonne Serve bedanke ich mich für die technischen Umsetzungen.

Sehr herzlich möchte ich mich auch bei allen Studenten bedanken, die im Rahmen von Studien- und Diplomarbeiten zum Gelingen der Arbeit beigetragen haben. Dies waren: Julian Bazan Aguirre, Funda Baysal, Tanja Bergen, Thomas Hivert, Steven Koenig, Carolin Prechtl, Rukmini Reddy, Nadiya Romanova, Manuela Saftescu, Iva Stoyanova und Atanas Trifonov.

Dem Bundesministerium für Bildung und Forschung (BMBF) danke ich für die finanzielle Unterstützung im Rahmen des Projekts „Aufreinigung von Biopolymeren mittels Elektrofiltration". Den Projektpartnern Herrn Dr. Gerald Brenner-Weiss, Frau Dr. Jelka Ondruschka und Herrn Dipl.-Ing. Frank Kirschhöfer möchte ich für die konstruktive und fruchtbare Zusammenarbeit danken.

Herrn Dipl.-Ing. Fritz Tiede von der Firma Optoprim GmbH möchte ich für die Anfertigung des Fluoreszenz-Systems und die sehr wertvollen Diskussionen danken.

Weiterhin möchte ich mich bei dem Deutschen Akademischen Austauschdienst (DAAD) für das Stipendium bedanken.

Ein besonders herzlicher Dank gilt auch meinen Eltern und meinem Bruder für die vielfältige Unterstützung.

Karlsruhe, Dezember 2011 Gözde Gözke

Contents

1 Introduction

1.1 Motivation

Biotechnology offers various major advantages in comparison to chemical synthesis. Amongst others these are the use of natural renewable feedstocks, the production of biodegradable and biocompatible final products, the moderate ambient reaction temperatures, and the formation of sequentially ordered, structured polymers. With the rapid growth in the field of biotechnology industry, processing of biopolymers which are generated from animal, agricultural or microbial sources have placed a premium on the development of their applications. However, a number of challenges predominantly in the isolation and purification part of bioprocesses need detailed consideration. A critical limiting factor in the development of a commercially significant biotechnological process is the downstream stage, which due to the increased number of post-fermentation operations could occupy up to 80% of the total production costs. Furthermore, the loss of product increases parallel with the number of downstream processing steps. The reason of high costs originates from the complexity of biosuspensions, the low concentration of the product, the multiplicity of byproducts, and the stress sensitivity of the objected molecules. An ideal bioseparation process step combines high separation efficiency and selectivity, with gentle process conditions and low production costs. Therefore, in order to develop a more efficient and economically feasible production procedure, it is necessary to reduce the number of purification steps, involved in downstream processing. Generally, membrane processes or precipitation/extraction techniques are the most common separation processes in biotechnology. However, precipitation of biopolymers by an organic solvent can lead to denaturation of the target product by binding to specific molecule locations and thus disrupting hydrophobic interactions which hold the structure together. Another disadvantage here is the necessity to recycle the precipitation agent, a process generally requiring a lot of energy. In this respect, membrane filtration processes have become an attractive alternative to many industrial applications due to the unique separation capabilities, and the easy capability of to scale up. Nevertheless, biopolymers are generally difficult to filtrate since they form together with biomass nearly impermeable, compressible filter cakes with high specific resistance. Possible mechanisms for compressibility include the collapse of solid particle bridges, and the concentration of fine particles in the liquid bridges of large particles. Hence, filtration technology of biopolymers is limited and in most cases classical dead-end filtrations cannot be applied for industrial scale bioseparations. In order to improve the performance, process selectivity must be increased.

Cross-flow filtration offers the hindrance of the formation of surface layer where the tangential flow over the membrane results in shear forces that reduce the thickness of the surface layer. The disadvantage of this process is that biopolymers, sensitive to shear stress, can be destroyed and high final product concentration cannot be achieved. In order to avoid limitations, innovative and integrative methods, based on the knowledge of the physicochemical properties of the biopolymers, need to be developed. The basic idea of such approaches is to apply additional forces to the particles such as an electrical force. Due to the dissociated residues and van der Waals forces, most bioparticles are charged and application of an electrical field leads to strong Coulomb forces acting on the particles. The ratio of particle charge to particle diameter is especially suitable for the mobility in an electric field.

Electrofiltration – the hybrid method which is the combination of membrane filtration and electrophoresis in a dead-end process is regarded as an appropriate method for concentration and fractionation of biopolymers. So far, the improvement and characterisation of electrofiltration was achieved successfully. However, there is still optimization potential for experiments with other product classes, studying filtration kinetics by variation of electrical field strength and/or pressure, and *in-situ* investigation of electrofiltration chamber regarding dynamics of the filter chamber system. This further improvement and knowledge provide a crucial opportunity for the commercial use of electrofiltration.

1.2 Aim of the work

The aim of the presented work is the purification and concentration of technically important biopolymers by means of electrofiltration and optimization of the process by investigating the *in-situ* behavior inside the filter chamber. The first part of the work states for the experiments which will be carried out on three model biopolymers – poly(3-hydroxybutyrate) (PHB) as a biopolyester, chitosan and hyaluronic acid as polysaccharides. These biopolymers are valuable prototypes in their product classes with wide range of application fields. The first product, PHB, is a biodegradable plastic which is produced in bacteria as intracellular granules and present a valuable alternative to petroleum based products with its ecological advantages. The second product, chitosan, is produced from skeletons of crustaceans or by fungi as cell wall component. The wide industrial applications range from water treatment to pharmaceutical fields. The third product, hyaluronic acid, is also produced by animal and microbial sources and has wide applications mostly in medical and cosmetic fields. After production of the substances in bioreactors, the application of electrofiltration will be investigated in the downstream processing step focusing on the effects of electric field strength and pressure on filtration kinetics. The analysis promotes an improvement and optimization in separation processes of the products. Improving filtration kinetics demonstrates that electrofiltration is an important and substitutional separation method alternative to the common methods which hedge about their separation. Testing new products with the improvement of process feasibility will open up new application areas for electrofiltration.

The second part of the work states for the optimization of electrofiltration system. Deeper understanding of the processes taking place inside the filter chamber leads to a valuable contribution for the process and apparatus optimization of electrofiltration. Regarding this, two new electrofiltration chambers will be developed. One of them is a filter chamber that will be constructed with fluorescence sensors which simulate the composition of the filter cake in the filter chamber by using a labeled biopolymer, FITC-dextran. Fluorescence as an investigative tool in studying the structure and dynamics of concentration gradient in filter chamber provides spatial and temporal information on the development of the electrofiltration system design. The other chamber is a new filter chamber that will be constructed with integrative voltage sensors which point the actual voltage strength in the filter chamber. The investigation on the magnitude of voltage drop on membranes and the distribution of the applied voltage between two electrodes will demonstrate useful information for the development of the system. Concerning the research, several different membranes will be tested to determine the optimum membrane which has relative lower voltage drop on membranes. On the other side of optimization experiments, the effect of conductivity in and out of the filter chamber will be characterized. The buffer solutions and the biopolymer solutions will be set with different conductivities to investigate the influence on filtration kinetics. Chromatographic analysis of the filtrate products will reveal information about the transported ions through the membranes. The experiments aimed at the effect, provide knowledge for the migration of ions through the membranes affecting the flow rate of the system and the membrane potentials.

To summarize, the aim of this work is to improve the electrofiltration system with an increased understanding of the factors involved in the process, leading new developments in downstream processing of biopolymers.

2 Theoretical background

2.1 Biopolymers

Biopolymers are classified according to their chemical structure into nucleic acids, polyamides, polysaccharides, polyesters, poly-isoprenoides and polyphenols. The primary structure of many biopolymers comprised of a linear chain of monomers. In this respect, they are similar to most synthetic polymers. However, they are produced by bacteria, fungi, plants and animals presenting specific behaviors (Kaplan 1998; Steinbüchel 2003).

Biopolymers contribute to the major fraction of the cellular dry matter and fulfill a wide range of essential or beneficial functions, including conservation and expression of genetic information, catalysis of reactions, storage of carbon, nitrogen, phosphorus and other nutrients and of energy, defense and protection against attacks by other cells or hazardous environmental or intrinsic factors, sensors of biotic and abiotic factors, communication with the environment and other organisms, mediators of the adhesion to surfaces of other organisms or non-living matter etc. In addition, many of them are structural components of cells, tissues and entire organisms (Steinbüchel and Doi 2005).

Among biopolymers, polysaccharides and polyesters exhibit fascinating properties and play a major role especially in the pharmaceutical, cosmetic and nutrition industries. Some technically relevant polysaccharides and polyesters are classified in Table 2-1.

Table 2-1: List of some important polysaccharides and polyesters (Kaplan 1998)

Polysaccharides			
Polysaccharides (plant/algal)	Polysaccharides (animal)	Polysaccharides (bacterial)	Polysaccharides (fungal)
Starch	Hyaluronic acid	Xanthan	Chitin
Cellulose	Chitin	Levan	Chitosan
Pectin	Chondroitin	Polygalactosamine	Pullulan
Konjac		Curdlan	Elsinan
Alginate		Gellan	Scleroglucan
Carrageenan		Dextran	
Gums		Hyaluronic acid	
Polyesters			
Polyhydroxyalkanoates, Polylactic acid, Polymalic acid			

Polysaccharides constitute one of the largest groups of natural compounds. They can be employed with a wide range of potential applications, as food additives due to their thickening and gel formation abilities, as renewable resources for the production of building blocks, as biofilms being an important component of the adhesive mechanism, as ligand systems using for the preparation of metal chelators in waste water treatment, as ion exchange resins and precious metal recovery, as separatory systems, as biosurfactants for the modification of interfacial and surface conditions, as bioactive materials in the protection and growth control of plants (Dumitriu 1998).

On the other side, most of the polyesters are synthesized by plants as structural components, or by prokaryotic microorganisms as intracellular storage components. The main applications are in the fields of packaging, bags, fibres, textiles, agricultural films, catering and fast-food, toys, leisure, medicine, hygiene and cosmetics and as bioplastics they present a high potential with respect to petrochemical supplies (Clarinval and Halleux 2005). Bioplastics can be bio-based but not biodegradable such as polyethylene and polyvinylchloride or bio-based and biodegradable such as polyhydroxyalkanoates. Biodegradable but not bio-based products such as polyvinyl alcohol and polycaprolactone are out of the bioplastics class.

Many bacteria are capable of synthesizing various types of polyesters including 3-,4-, and 5- hydroxyalkanoate units. Approximately 150 different hydroxyalkonates are known as constituents of these polyesters. Therefore, they are generally referred to as polyhydroxyalkanoates. The PHA family of polyesters is thermoplastic with biodegradable and biocompatible properties leading an alternative use to petrochemical thermoplastics. By varying the composition of the polyesters, the physical and mechanical properties can be regulated. As a result, PHAs has a wide variety of applications from hard crystalline plastics to very elastic rubber (Doi and Steinbüchel 2002).

Interdisciplinary research and development of bioplastics is a rapidly growing field. The global production capacity of bioplastics, which are generally polyesters, increases continuously from 262000 tons world-wide in 2007 to an estimated 1502000 tons in 2011 (see Figure 2-1).

This tendency represents the market potential obviously. The market demand for biopolymers is strongly influenced by many factors such as competitive price, different applications, legislative laws and optimization of commercial composting process (European Bioplastics 2007).

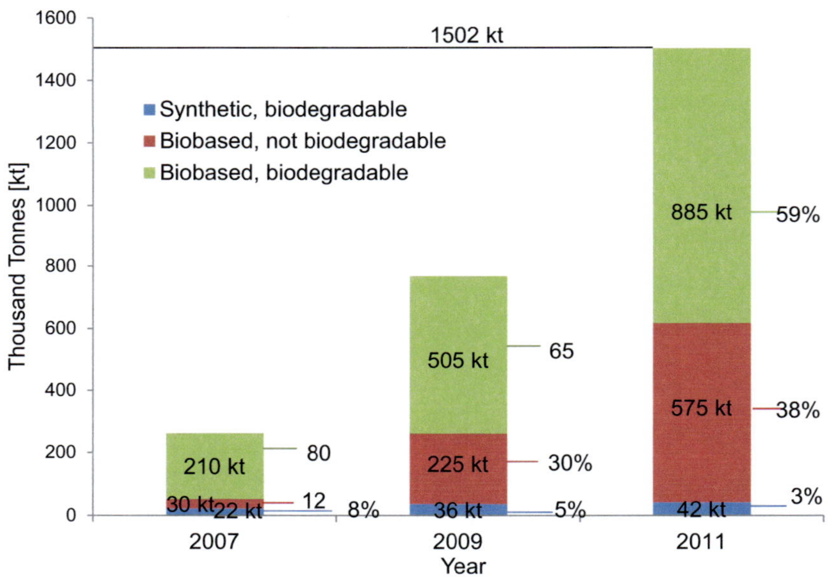

Figure 2-1: Estimated annual bioplastics production capacity 2007-2011 (European Bioplastics 2007)

Due to their attractive properties and various applications, biopolymers emerge as an important component for economic development. Production process generally consists of three main steps. Downstream processing is followed after upstream processing (genetic engineering, media optimization, biocatalyst screening) and bioreaction part (fermentation, cell culture or enzymatic reaction). Downstream processing refers to recovery, isolation, purification and polishing of the product. However, one of the major problems for the development of biopolymers is related to their cost which mostly arises from the downstream processing steps such as cell disruption, separation, and purification (Ghosh 2006).

Bioseparation techniques have to be gentle in terms of avoiding extremes of physicochemical conditions such as pH and ionic strength, hydrodynamic conditions such as high shear rates. Organic solvents which are widely used in chemical separations have relatively limited usage in bioseparations due to their tendency to promote degradation of many biological products (Ghosh 2006).

PHB, hyaluronic acid and chitosan are three of the widely used technically important biopolymers with remarkable properties and the large variety of potential applications. Nevertheless, the interest on these biopolymers is accompanied by the difficulties in downstream processes which need an intensive analysis in order to expand the market.

2.1.1 Poly(3-hydroxybutyrate) (PHB)

The last two decades are marked by the ever growing environmental concern with echoes in society and more importantly in industry. This is accompanied by production profile changes and invention of new recycling/disposal strategies. The disposal of plastics is already an existing worldwide environmental problem. Hazards and problems are derived from the stability of the plastics, persisting with years, once released in nature, threatening the balance in the whole ecosystem. Incineration and post-consumer recycling are some of the strategies adopted in many countries. The permanent and global solution of the problem emerges from the development of biodegradable polymers. As an alternative to petroleum-related plastics, new products based on renewable raw materials become increasingly important. The variety of special applications, define PHB, as relevant, desired and attractive product in the current market frame (Koller et al. 2008). Many companies are involved in the production of PHB (Reddy et al. 2003). However, commercial applications of PHB are hindered due to the high costs which mainly originate from substrate (Poirier et al. 1995) and downstream processing costs (Reddy et al. 2003).

PHB is the most common member of the polyhdroxyalkonates (PHA) family. It is a polyester which belongs to a short chain PHA with monomers having 4-5 carbon atoms (Doi and Steinbüchel 2002). The chemical structure of PHB is presented on Figure 2-2.

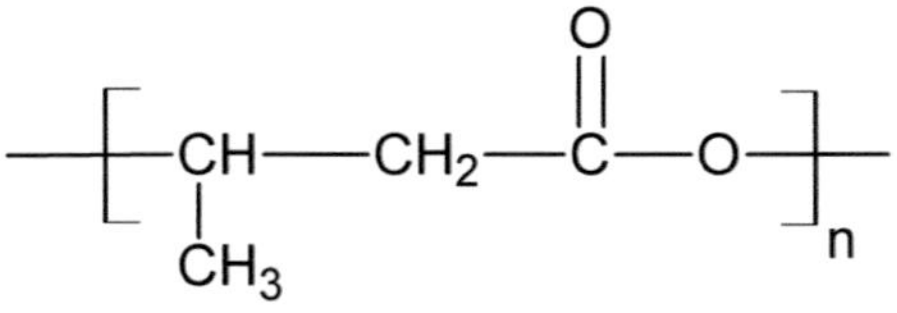

$$\left[\begin{array}{c} CH \\ | \\ CH_3 \end{array} - CH_2 - \overset{\displaystyle O}{\overset{\|}{C}} - O \right]_n$$

Figure 2-2: Chemical structure of PHB

Many different bacteria species synthesize intracellularly PHB as carbon and energy storage material (Chen 2005; Hazer and Steinbüchel 2007), if subjected to nutrient limiting conditions with a sufficient supply of the carbon source (Anderson and Dawes 1990). Among PHB producing bacteria *Ralstonia eutropha* is the most extensively studied one due to its high accumulation capacity resulting in PHB formation up to 80% of the cell dry weight (Tavares et al. 2004). PHB can also be synthesized in plants, for instance Arabidopsis thaliana (Poirier et al. 1992; Valentin et al. 1999), accumulated with low yield in the cytoplasm of tobacco (Nakashita et al. 1999) and cotton fiber cells (John and Keller 1996), being the major problem for its plant production (Slater et al. 1999).

PHB is a biodegradable, biocompatible, piezoelectric and thermoplastic biopolymer which has physical and mechanical properties similar to polypropylene (Madison and Huisman 1999) with wide range of application areas, including medicine, agriculture and packaging industry (Patnaik 2005). In addition, PHB is more crystalline, more brittle and denser than polypropylene (Wang et al. 2005), which together with the biodegradability make it preferred and attractive polymer. The biodegradation occurs in natural environments and PHB can be metabolized to CO_2 and H_2O through an aerobic bacterial pathway (Imam et al. 1999; Jendrossek and Handrick 2002). PHB is not soluble in water and oppositely to most biopolymers, relatively resistant to hydrolysis (Clarinval and Halleux 2005). PHB granules consist of approximately 97 to 98% pure PHB, 2% protein and 0.5% lipids (Kawaguchi and Doi 1990). The protein film on these granules contributes to the negative charge of the biopolymer although as a rule PHB has a low charge.

In the downstream processing of PHB, the major step is the extraction step which involves the disruption of the cells by chemical, enzymatic or mechanical methods (Jacquel et al. 2008). Figure 2-3 presents the technological steps in both available and already commercially applied PHB purification scheme and the novel approach by electrofiltration. In order to increase the effectiveness of the extraction step, some chemical or physical pretreatments can be applied such as heat, alkaline, salt or freezing treatments leading also the increase in the production costs (Fiorese et al. 2009). The extraction process usually involves the use of organic solvents such as chloroform, methylene chloride, 1,2-dichloroethane, or pyridine. The solution is filtered to eliminate bacterial cell debris and then the biopolymer is precipitated by adding to a non-

14

solvent such as methanol, diethyl ether, or hexane (Hocking and Marchessault 1998) or the solvent is evaporated (Mothes et al. 2007). For the purification, PHB is redissolved in chloroform and reprecipitated with hexane or diethyl ether. (Hocking and Marchessault 1998). Addition of large amounts of organic solvents is hazardous to the environment and increases the total cost (Khosravi-Darani et al. 2004). Moreover, in the case of evaporation is applied, energy need increases.

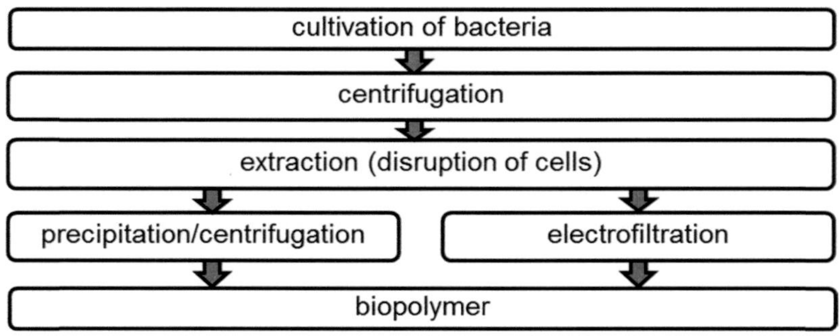

Figure 2-3: Traditional commercial and novel electrofiltration purification of PHB dispersion

Enzymatic digestion method is an attractive choice due to its high efficiency. However, the application of enzymes is an intricate procedure. Mechanical cell disruption is another method that can be applied before purification step (Tamer et al. 1998) and needs several passes.

In order to increase the level of purity of the biopolymer hydrogen peroxide treatment combined with enzymes or chelating agents can be applied. Nevertheless, hydrogen peroxide treatment of PHB has many drawbacks such as high operating temperatures (80-100°C), an instability of peroxides in the presence of high levels of cell biomass and a decrease in the molecular weight of the biopolymer (Jacquel et al. 2008).

Electrofiltration method as a novel purification strategy can be applied after extraction step. The granules coated with protein film, exhibit relatively high zeta potential, enabling purification of PHB by means of an electric field. The ease of application of electrofiltration corresponds to the demand for cost reductions and higher product recovery in the downstream processing.

2.1.2 Chitosan

Chitin is the second most abundant natural polysaccharide after cellulose though as a result of poor solubility, its technical applications are limited (Rinaudo 2006). These technical obstacles are avoided by introducing chitosan – a chitin derivative produced by alkaline deacetylation, which is completely soluble in acidic aqueous solutions. The structures of chitin and chitosan are presented on Figure 2-4.

The derivative chitosan is a heteropolymer composed of glucosamine (2-amino-2-deoxy-β-D-glucopyranose) and acetylglucosamine (2-acet-amido-2-deoxy-β-D-glucopyranose) residues. An important parameter characterizing chitosan is the degree of deacetylation (D_{DA}), defining the deacetylated percentage of chitin. When D_{DA} is ≥50%, the investigated product is designated as chitosan (Khor 2003). The majority of the commercially available chitosan features D_{DA} ranging normally between 70-90% which affects the physical, chemical and biological properties (Jiang et al. 2003; Balazs and Sipos 2007).

Chitosan is a biocompatible and biodegradable polysaccharide with applications in food and flavor industry, photography, cosmetics, pharmaceutical products for wound healing, ophthalmology, water engineering, paper manufacture industry, drug delivery systems etc. (Kumar 2000). The interest in this product is followed by the necessity for improvement of its downstream processing, due to the time consuming operations and the possible structural changes during purification, which decrease the final quality.

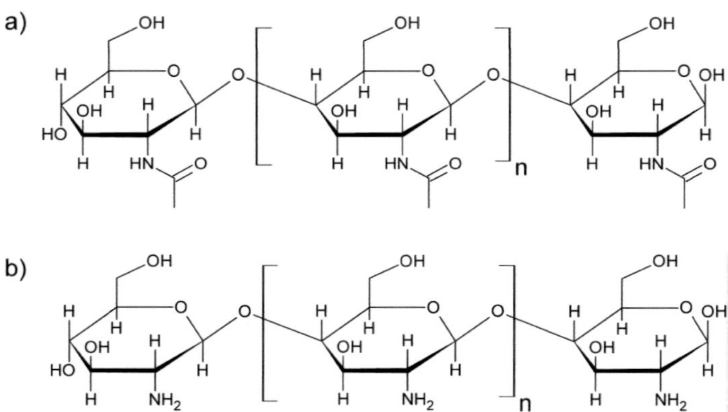

Figure 2-4: Chemical structure of chitin (a) and chitosan (b)

Chitosan can be derived from cell walls of fungi or from exoskeletons of arthropods (Ondruschka et al. 2008). The different origin also determines different properties and, respectively, fields of application. The advantages of chitosan with fungi origin feature all-year-long character of the production instead of seasonal one, and the heavy metal free character of the product, affecting its properties and qualifying it as attractive for medical applications (Chung et al. 2004). In addition, fungi derived chitosan has lower molecular weight which is beneficial for medical applications (Groeger et al. 2006; Yang et al. 2009).

The production process includes the same technological operations for fungus (Rane and Hoover 1993; Zamani et al. 2007; Ondruschka et al.

2008) and for crab derived chitosan (Teng 2001; Yen et al. 2009). It includes acid extraction or enzymatic hydrolysis in the presence of chitin deacetylase (Gao et al. 1995; Cai et al. 2006). Chitosan is synthesized directly as cell wall component by species from order *Mucorales* like *Absidia coerulea, Mucor rouxii* and *Rhizopus oryzae*, facilitating the necessity of deacetylation step in the production technology (Rane and Hoover 1993; Ondruschka et al. 2008).

Chitin from both fungi and arthropods is captured within a protein net, which has to be removed in order to obtain a high quality pure product. Although some fungus-derived chitin preparations exhibit lower protein content, the promising medical applications require a higher level of purity advanced by further purification steps (Groeger et al. 2006).

The common chitosan production process includes several steps: deprotonation, demineralization, deacetylation, acid extraction, pre-cipitation, and centrifugation. Chemical deacetylation of chitin to chitosan is achieved by treatment with aqueous 40-50% NaOH at 110-130°C for several hours (Trutnau et al. 2009). However, the use of Zygomycetes species *Absidia coerulea* and *Mucor rouxii* facilitate this step, because their cell walls already contain chitosan in addition to chitin.

The presence of free amino groups in the structure of chitosan generates a positively charged molecule leading to a high zeta potential and enables its purification by electrofiltration. The alternative pathway in this procedure applies electrofiltration instead of step precipitation and afterwards centrifugation. The technological steps in the production of chitosan are presented on Figure 2-5.

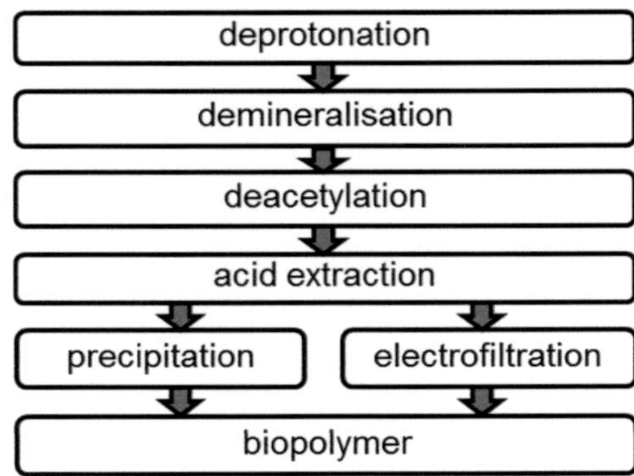

Figure 2-5: Technology scheme for production of chitosan derived from fungi and crustaceans

Electrofiltration reduces production time and in addition avoids the use of concentrated NaOH solution during the precipitation step, and thus reduces the generation of a large amount of concentrated alkaline solution which can give rise to waste disposal problems and lead to structural changes of the product.

2.1.3 Hyaluronic acid

Hyaluronic acid (HA) is a linear polysaccharide which is comprised of alternating glucuronic acid (GlcUA) and N-acetylglucosamine (GlcNAc) units with a molecular weight in the range of $1\text{-}10^4$ kDa depending on its source and production process (Kim et al. 2006). It was discovered by Meyer and Palmer in 1934 in the cattle vitreous humor (Meyer and Palmer 1934). Due to the polyanionic form, it is also designated as hyaluronan (Kakehi et al. 2003). Figure 2-6 presents the chemical structure of hyaluronic acid. The physicochemical properties of hyaluronic acid such as biodegradability, hydrophilicity and bio-compatibility make it an excellent biomaterial with a wide variety of medical and cosmetic applications such as wound healing, skin moisturizing, and osteoarthritis treatment (Andre 2004; Yoo et al. 2005; Chong et al. 2005).

Hyaluronic acid in the form of a hydrated gel is found in human and animal tissues such as the vitreous humor of the eyeballs, the synovial fluid, and also contributes for the formation of skin layers. In addition, certain bacteria have the ability to synthesize HA as part of their outer capsule (Ignatova and Gurov 1990) and amongst them groups A and C *Streptococcus* sp. have commercial potential (Blank et al. 2005; Kim et al. 2006; Duan et al. 2008; Chen et al. 2009). Besides streptococcal bacteria, genetically modified Bacillus subtilis strains are recently tested for their ability to produce hyaluronic acid (Ogrodowski et al. 2005; Chien and Lee 2007).

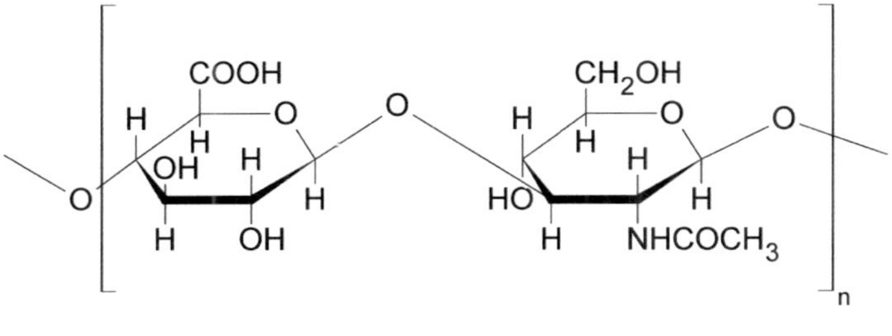

Figure 2-6: Chemical structure of hyaluronic acid

Alternatively and with historical significance HA is produced from rooster combs by extraction (Boas 1949; Nakano et al. 1994; Sousa et al. 2009). Technological schemes for the production of hyaluronic acid are presented on Figure 2-7. However all rooster comb based HA products carry warnings to people, allergic to avian products. There are already reports for inflammatory reactions after injection of such products (Noble 1993). Another disadvantage arises from the difficulties in isolation, for

HA easily forms complexes with proteoglycans during extraction (O`Regan et al. 1994). Therefore more efforts and researches are directed towards the bacteria based synthetic pathway. The biotechnological production of HA by *Streptococcus* sp. provides better control over molecular weight distribution, purity of the product and excludes the use of solvents in the purification stages.

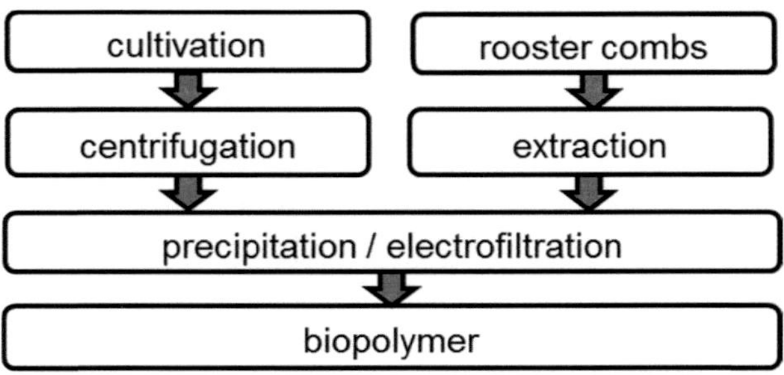

Figure 2-7: Production of hyaluronic acid

The fermented broth is usually centrifuged in order to remove cells and debris (Sousa et al. 2009). As an alternative method to centrifugation, cross-flow filtration was previously investigated (Zhou et al. 2006). Purification of HA mainly performed by solvent precipitation (Sousa et al. 2009). Nevertheless this time and energy consuming process additionally increases costs. Many of the commercial applications depend on the molecular weight of hyaluronic acid, for example cosmetic industry needs predominantly lower molecular weight molecules, while pharmaceutical industry, on the contrary. Therefore, this is an important parameter, needing to be considered during synthesis.

In biotechnology the current downstream processing trends are directed towards integrated, faster and more effective processes, incorporating mild operational conditions. Different approaches including membrane filtration are introduced to solve the technological challenges. The electric charge of HA enables the application of electrophoretic separation (Kakehi et al. 2003; Hutterer and Jorgenson 2005).

Despite the serious number of advantages, introduced by the production of hyaluronic acid by microorganism, this synthetic pathway is still in its early research stages. The full potential of this approach will be revealed after precise optimization of the process, leading to product with sufficient purity, defined properties and acceptable self-value.

2.2 Interfacial phenomena

Difficulties in the filtration of biopolymers generally arise from the high viscosities and physicochemical interactions within filter cakes, controlling permeation rates. These interactions depend on London-van der Waals forces, hydration forces, electrostatic (double layer) forces and configurational entropic effects (steric effects) (Medronho 2003).

For the filtration experiments with BSA, (Bowen and Williams 1996) confirmed that the configurational entropic effect becomes quite important with the decrease of the colloid size. This effect can be neglected for macromolecules like polysaccharides whose molecular weights are 10-100 times bigger than proteins (Hofmann and Posten 2003).

Electrostatic and London-van der Waals interactions are the main effects, forming together the basis of DLVO theory (Dörfler 2002). The scientists Derjaguin, Landau, Verwy and Overbeek developed in the 1940s a theory, describing the stability of colloidal systems (Derjaguin and Landau 1941, Verwey and Overbeek 1948). The basic idea of the theory is that the stability of dispersion is determined by the sum of attractive and repulsive forces between individual particles. The DLVO theory for colloidal interactions postulates that a colloidal system will remain stable only when the electrostatic repulsion is greater than the van der Waals force between particles (Dörfler 2002). Biological colloids behave in a similar fashion in aqueous solution as other types of colloidal particles (García et al. 1999).

The attraction between the particles arises from dispersion forces, often called London-van der Waals forces. The London interaction is a consequence of charge fluctuations within an atom or molecule associated with the motion of its electrons. Other van der Waals forces such as dipole-dipole (Keesom), dipole-induced dipole (Debye) can be neglected for colloidal particle interactions. The van der Waals energy of attraction between two equal colloidal particles each of radius a, at a distance h in vacuum, is given by the equation 2.1 (Tadros 2007):

$$G_A = -\frac{A_H}{6}\left[\frac{2}{s^2-4}+\frac{2}{s^2}+ln\left(\frac{s^2-4}{s^2}\right)\right] \qquad 2.1$$

where $s = (2a+h)/a$ and A_H is the Hamaker constant. For very short distances of separation ($h<<R$) it can be approximated by:

$$G_A = -\frac{aA_H}{12h} \qquad 2.2$$

Another type of molecular interaction is electrostatic interactions which occur due to the positive or negative charges of the macromolecules emerging in aqueous electrolytes. These charges appear due either to preferential adsorption of certain ions from solution or to ionization of groups at the particle surface and they increase electrostatic repulsions in solution (Medronho 2003). In general, biopolymers are usually positively or negatively charged as a result of their functional groups, which open up vast new possibilities for the purification of biotechnological products. These charged biopolymers are polyelectrolytes bearing ionic groups such as $-COO^-$ (for hyaluronic acid), $-NH_3^+$ (for chitosan in acidic media). Electrostatic effects play an essential role in the interactions between many biologically important molecules (Nakamura 1996). Moreover, these effects have an important influence on the cake voidage and the specific filter cake resistance for proteins (Iritani et al. 1995; Bowen and Williams 1996) and for other biopolymers such as polysaccharides and polyesters (Gözke and Posten 2010).

The electroneutrality of the system requires that the charged surface be surrounded by oppositely charged ions, called counter ions (Kissa 1999). This fact leads to the formation of electrical double layer which comprises the particle charge and the counter charge together. The repulsion between particles arises from the interaction of the electrical double layers surrounding each particle. The electrical double layer makes an important contribution to the stability of colloidal dispersions and provides an explanation for the effect of added electrolytes on the stability of colloids (Barnes and Gentle 2005). The thickness of the electrical double layer around the particle, which is given by the Debye length $1/\kappa$ (κ: Debye-Hückel parameter), determines not only the decay distance of the electrical potential but also that of the distribution of the liquid velocity around the particle moving in an electrolyte solution (Ohshima 2006). For a general electrolyte composed of N ionic mobile species of valance z_i and bulk concentration (number density) n_i^∞, the Debye-Hückel parameter κ is defined by equation 2.3:

$$\kappa = \left(\frac{1}{\varepsilon_r \varepsilon_0 kT} \sum_{i=1}^{N} z_i^2 \, e^2 n_i^\infty \right)^{1/2} \qquad\qquad 2.3$$

where ε_r is the relative permittivity of the electrolyte solution, ε_0 is the permittivity of vacuum, e is the elementary electric charge, k is the Boltzmann constant, T is the absolute temperature.

Due to the release of counter ions by dissociation, ionic strength of solution and therefore the Debye length is influenced. The equation 2.3 indicates that the parameter $1/\kappa$ increases with decrease in the valency of the ions and decrease in electrolyte concentration.

Several models have been introduced for the structure of electrical double layer. The simplest model is that of Helmholtz: ions of charge opposite to the charge in the solid surface adsorb on the solid and completely neutralise its charge. The inner layer is made of dehydrated anions from the outer hydrated cations. The potential is drastically reduced with increasing distance from the surface linearly to zero. This model is unrealistic because it ignores the thermal motion of the ions and the solvent molecules. The Gouy-Chapman theory describes a diffuse layer of ions in which the charge on the surface is progressively neutralised by an excess of oppositely charged ions away from the surface (Barnes and Gentle 2005). In this model, there is no fixed connection between particle and ion cloud. The model of Stern is a combination of older models by Helmholtz and Gouy-Chapman. Stern proposed a model in which the double layer is divided into two parts separated by a plane (the Stern plane) located at about a hydrated ion radius from the surface, and also considered the possibility of specific ion adsorption. Specifically adsorbed ions are those which are attached to the surface by electrostatic and/or van der Waals forces strongly enough to overcome thermal agitation. The centers of any specifically adsorbed ions are located in the Stern layer. Ions located beyond the Stern plane form the diffuse part of the double layer. The potential changes from ψ_0 (the surface potential) to ψ_d (the Stern potential) in the Stern layer, and decays from ψ_d to zero in the diffuse double layer (see Figure 2-8).

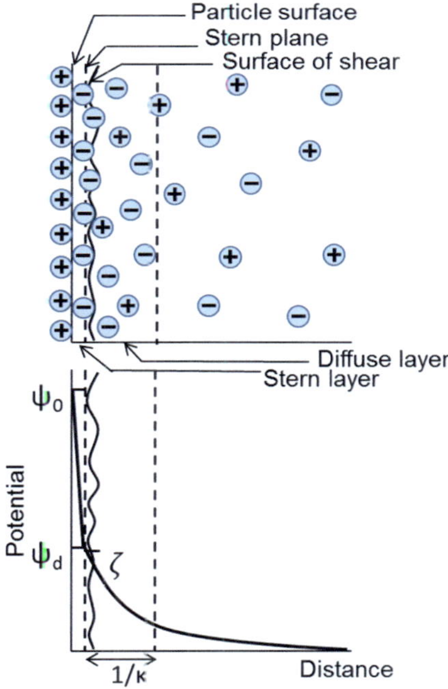

Figure 2-8: Schematic representation of the structure of the electrical double layer according to Stern's theory (Shaw 1992, Harrison et al. 2003)

Shear plane is the boundary between the Stern and Gouy-Chapman layers. The potential at the surface of shear between the charged surface and the electrolyte solution is called zeta potential (ζ) and can be estimated from electrokinetic measurements. In addition to ions in the Stern layer, a certain amount of solvent will probably be bound to the charged surface and form a part of the electrokinetic unit. It is therefore reasonable to suppose that the shear plane is located at a small distance further out from the surface than the Stern plane and that ζ is, in general, marginally smaller in magnitude than ψ_d (Shaw 1992).

The magnitude of zeta potential is correlated to the colloid stability due to the mutual repulsion. The electrostatic energy of repulsion, G_{el}, is given by the following expression which is valid for $\kappa.a<3$ (Tadros 2007):

$$G_{el} = \frac{4\pi\varepsilon_r\varepsilon_0 a^2\psi_d^2 exp(-\kappa h)}{2a + h} \qquad 2.4$$

Equation 2.4 shows that G_{el} decays exponentially with increase of h and it approaches zero at large h. The rate of decrease of G_{el} with increase in h depends on $1/\kappa$: the higher the value of $1/\kappa$, the slower is the decay. In other words, at any given h, G_{el} increases with increase in $1/\kappa$, i.e. with decrease in electrolyte concentration and valency of the ions (Tadros 2007).

Degree of charge is affected by the nature of the polymer, pH and ionic strength of the polyelectrolyte solution (Kissa 1999). These modifications in structure lead to change in filtration kinetics. The pH of solution and the electrolyte strength play an important role in filtration's behavior, affecting the porosity of filter cakes of binary protein mixtures (Iritani 2003). Iritani et al. (2008) experimentally determined the filtration characteristics – average porosity and average specific filtration resistance, for yeast suspensions. In addition, filter cake porosity can be simulated, considering particle size, compressibility of the cake and the physicochemical behavior of the system (Neesse et al. 2009). The electrostatic influence on particles varies with the change in zeta potential, caused by different pH and ionic strength (Lockhart 1992; Orsat et al. 1999). Various scientists investigated the effects of zeta potential, pH value and ionic strength on the recombinant protein lactoferrin (Bowen et al. 1999), on mixtures of BSA and lysozyme (Iritani et al. 1995) and on mixtures of rutile titanium dioxide (TiO_2) and BSA (Mukai et al. 2009). Similar results have been obtained for the ionic strength and pH dependence of BSA in cross-flow ultrafiltrations (Opong and Zydney 1991). Important are the interactions not only between polymers but also between polymer and the filtration membrane. The intermolecular forces between lysozym and polyethersulfon membrane were investigated by Koehler et al. (1997), providing the molecular basis for the understanding the membrane fouling, resulting from the filtration of protein solutions with polymeric membranes. Experiments confirmed that both lysozyme-lysozyme and lysozyme-polyethersulfone interactions are equally responsible for the decline in filtration performance. For further better understanding of the influence of protein and membrane type interactions upon filtration, the fouling effects, when using two different kinds of polymeric membranes, with different surface roughness and chemistry, were investigated. The results obtained demonstrated that hydrophilic membranes are less subjected to fouling than the hydrophobic ones (Bowen et al. 2002).

2.3   Electrokinetic phenomena

Electrokinetic effects in dispersions are associated to the surface charge and the electric double layer of the dispersed particles. Concerning electrofiltration, the movement of a charged particle can cause some electrokinetic phenomena such as electrophoresis, electroosmosis, and streaming potential. These electrokinetic phenomena differ from each other in the type of applied external field and the measured quantity (Table 2-2).

Table 2-2: Schematic definitions of the main electrokinetic phenomena that introduced in electrofiltration (Delgado and Shilov 2006)

Type of external field	Measured quantity	Name
Electric	Particle mobility	Electrophoresis
Electric	Liquid flow rate	Electroosmosis
Pressure gradient	Electrical potential difference	Streaming potential

2.3.1 Electrophoresis

Electrophoresis is the movement of a dispersed particle and its electric double layer by an external electric field. When an electric field is applied to colloidal particles suspended in an electrolyte solution, the mobility of a particle, u is generally defined as the velocity of the particle, v, per unit electric field, E (Caplan et al. 1995):

$$u = \frac{v}{E} \qquad\qquad 2.5$$

The external field accelerates the particles and at the same time a viscous force exerted by the liquid on the particles tends to retard the particles (Ohshima 2006). Therefore, two forces with opposite directions take effect – the hydrodynamic resistance force which is defined by Stokes law (F_W) (Hamann et al. 2007) and the electrophoretic force (F_E) which is defined by Coulomb's law (Hamann et al. 2007). These forces are expressed as in given in equations 2.6 and 2.7 respectively:

$$F_W = 6\pi v a \eta \qquad\qquad 2.6$$

$$F_E = qE \qquad\qquad 2.7$$

For steady state conditions charged colloids have constant migration velocity and the sum of all the forces is zero (equation 2.8). When all other effects are ignored the consideration of electrophoretic force and hydrodynamic resistance force results in equation 2.9:

$$\sum F = 0 \qquad\qquad 2.8$$

$$qE = 6\pi va\eta \qquad\qquad 2.9$$

It is seen that the velocity of the particles is directly proportional to the electric field (E) and the net charge of the particle (q). Substitution of equation 2.10 to 2.9 gives equation 2.11 (Hunter 1993):

$$\zeta = \frac{q}{4\pi\varepsilon a} \qquad\qquad 2.10$$

$$u = \frac{2\varepsilon\zeta}{3\eta} \qquad\qquad 2.11$$

Electrophoresis depends strongly on the electrical double layer formed around the particles and the zeta potential. Together with electric field strength, dielectric constant and viscosity of the medium, zeta potential is an important factor that affects the velocity, respectively the electrophoresis of the particle. Its magnitude is correlated to the colloid stability due to the mutual repulsion (Dörfler 2002). Increasing the ionic strength and thereby decreasing the double layer thickness produces a decrease in the zeta potential, because its position is dictated by hydrodynamic effects involving shear plane (García et al. 1999).

A parameter of frequent use and of high significance for the study of equilibrium and kinetic colloidal phenomena, connected with the electric charge of colloidal particles is the dimensionless parameter κa, the ratio of the characteristic radius of curvature of particle's surface to the thickness of the diffuse layer. Henry took into account the effect of the particle shape and size on the electric field and formulated the equation 2.11 into equation 2.12 (González-Caballero and Shilov 2006):

$$u = \frac{2\varepsilon\zeta f(\kappa a)}{3\eta} \qquad\qquad 2.12$$

Limiting cases of very small and very large values of κa parameter presents characteristic simplifications. For thick double layers ($\kappa a \ll 1$) in low dielectric constant media $f(\kappa a)$ becomes 1.0 and this equation is referred to Hückel equation. For thin double layers ($\kappa a \gg 1$) in aqueous media $f(\kappa a)$ becomes 1.5 and this equation is referred to Smoluchowski equation. Hückel's equation differs from Smoluchowski's equation by a factor of 2/3 (Ohshima 2006). For longer cells and vesicles (biological cells and large lipid vesicles) the Smoluchowski equation is often used and for small biocolloid particles (such as lipid micelles) the Hückel equation (Caplan et al.1995).

In electric field particles and their oppositely charged ion atmosphere migrate in opposite directions. As a result deformation of the ion atmosphere occurs (Weber and Stahl 2002). This effect causing the slow-down of the particle is known as relaxation effect (F_L) (see Figure 2-9). By ions migration another effect also takes place. Ion atmosphere draws together hydrate sheath of solvent. The motion of the ion atmosphere with the hydrate-sheath of solvent leads to a hydrodynamic backflow on the particle. This effect is known as retardation effect (F_T) (see Figure 2-9), and describes the migration rate of the particle with relaxation effect (Weber and Stahl 2002). These effects are ignorable when the particle diameter is significantly smaller or larger than the thickness of the double-layer, which consists of oppositely charged ions atmospheres, surrounding the polyion. They are considered insignificant for small colloidal ions and low ionic strengths and also for big ions and high ionic strengths (Schwuger 1996).

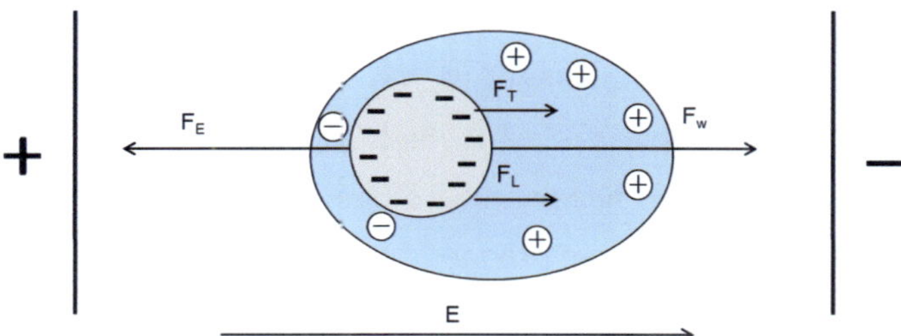

Figure 2-9: Macroions in an electric field

The electrophoretic mobility of linear polyelectrolytes in aqueous dispersions is described by Manning Theory (Manning 1981). This theory is valid when the interpolymer interactions are not taken into consideration, therefore only for high diluted polymer dispersions. It was applied in practice by Jones et al. (1987) in experiments with the polymer xanthan. Hofmann and Posten applied a model based on Manning Theory (Hofmann and Posten 2003). With this model filtration kinetics, respectively filtrate flow and the energy consumption of electrofiltration was simulated.

Mobility of an ion in an electric field

The charge of a simple ion can be written as the product of its valance z times the electron charge e (Hiemenz and Rajagopalan 1997):

$$q = ze \qquad \text{2 13}$$

Substitution of this result into equation 2.9 gives:

$$u = \frac{zeE}{6\pi\eta a} \qquad \text{2 14}$$

Mobilities are typically on the order of 10^{-8} m^2/Vs for simple ions (Hiemenz and Rajagopalan 1997).

2.3.2 Electroosmosis

Electroosmosis is the movement of the liquid adjacent to a charged surface under an applied electric field (Delgado and Shilov 2006). Electric field induces the acceleration of the ions in diffuse layer and the liquid adjacent to the ions are dragged by viscous effects. The velocity of the solution flow is described by the Smoluchowski equation (Kornyshew et al. 2002):

$$u = \frac{\varepsilon \zeta}{\eta} \qquad \qquad 2.15$$

where η the dynamic viscosity of the solvent and ε is the relative dielectric permittivity of the solvent.

2.3.3 Streaming potential

An inverse generation of an electric field inside the membrane takes place if the solution passes through this porous medium due to an imposed hydrostatic pressure. The phenomenon is called streaming potential. The flow of the liquid inside the pores induces the displacement of the mobile part of the electrical double layer at the surface of membrane pores, with respect to the charges attached to the surface. These dipoles create an electric field, which, under stationary conditions, prevents the farther displacement of the mobile charges. The resulting potential difference across the membrane, $\Delta\psi$, is proportional to the excessive hydrostatic pressure ΔP (Kornyshew et al. 2002):

$$\frac{\Delta\psi}{\Delta P} = \frac{\varepsilon\zeta}{\eta\Lambda}$$

2.16

where Λ is the electrolyte conductivity.

2.4 Principles of electrofiltration

2.4.1 Filtration

Membrane separations usually have low energy requirements compared with processes such as distillation, crystallization or evaporation wh ch involve significant energy inputs. The conditions under which the separations are conducted are normally mild, which is an important consideration when complex heat sensitive materials are used. The low energy requirements and mild operating conditions are among the principal reasons why memorane processes have been chosen for a range of biotechnological separations (Bell and Cousins 1994).

Filtration is an operation that has found an important place in the downstream processing of biotechnology products. In general, filtraton is used to separate particulate or solute components in a fluid suspension or solution according to their size by flowing under a pressure differential through a porous medium (Harrison et al. 2003).

Filtration in biotechnology can be applied in many different stages of downstream part of the process such as separation of cells, concentration of the product solution, sterile filtration of biopharmaceutical products, and filtration of precipitated substances (Harrison et al. 2003).

There are two broad categories of filtration, which differ according to the direction of the fluid feed in relation to the filter medium. In conventional filtration (dead-end filtration), the fluid flows perpendicular to the medium, which generally results in a cake of solids depositing on the filter medium. As the thickness of the cake increases, the permeate flux decreases with time, ultimately reaching zero. In cross-flow filtration (tangential flow filtration), the fluid flows parallel to the medium to minimize buildup of solids on the medium. The permeate flux reaches a constant value at steady state. The principle of conventional filtration and cross-flow filtration are illustrated in Figure 2-10 (Harrison et al. 2003).

In conventional filtration, due to the depositing of the filter cake on the membrane, the pores of the membranes can easily be blocked preventing an efficient operation (Gözke and Posten 2010).

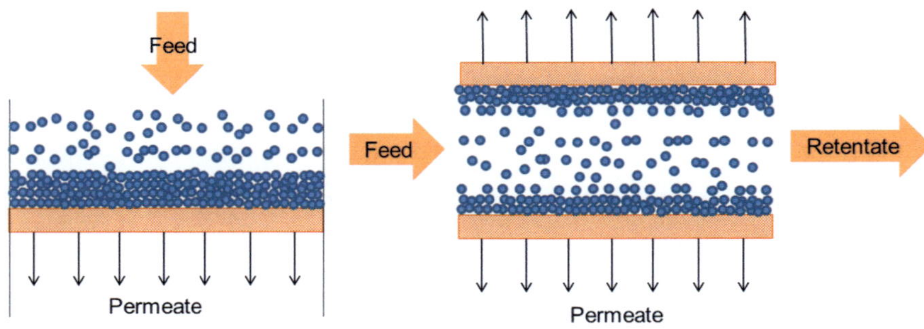

Figure 2-10: Schematic diagrams for dead and or conventional filtration (left) and cross-flow filtration (right) (Harrison et al. 2003)

In cross-flow filtration the tangential flow over the membrane results in shear forces that reduce the thickness of the surface layer. The disadvantages of this process are that biopolymers sensitive to shear stress can be destroyed and that high final concentrations cannot be achieved (Russotti and Göklen 2001).

A number of membrane processes have evolved, which uses a pressure driving force and a semi-permeable membrane in order to effect a separation of components in a solution or colloidal dispersion. According to molecular size pressure driven membrane processes mainly can be classified into microfiltration (MF), ultrafiltration (UF), nanofiltration (NF) and reverse osmosis (RO) separations. The dimensions of the components involved in these separations are given in Figure 2-11, and are typically in the range of less than 1 nm to over 1000 nm (Lewis 1996).

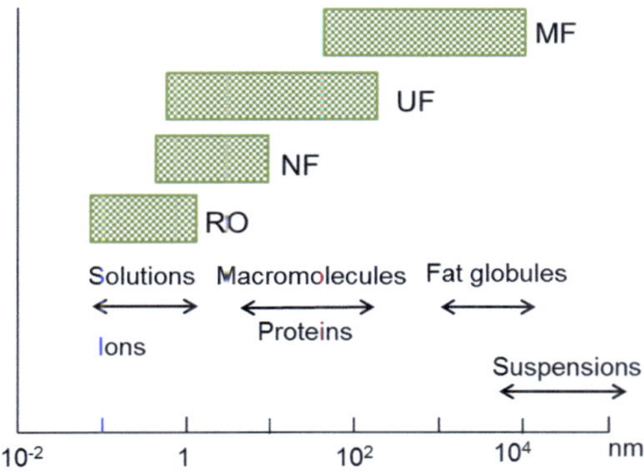

Figure 2-11: Size ranges for different membrane processes (Lewis 1996)

The model of cake filtration, based on Darcy's law (Darcy 1856, Anlauf 1994), has been applied for the description of polymer dead-end filtration. Darcy's law can be formulated with equation 2.17, where V is filtrate volume, t_f is filtration time, Δp_H is hydraulic pressure, A is filtration surface area, η is dynamic viscosity, α_{av} is average specific filter cake resistance, $m_{cake,dry}$ is dry mass of filter cake:

$$\frac{dV_f}{dt_f} = \frac{\Delta p_H A}{\eta \left(\alpha_{av} \dfrac{m_{cake,dry}}{A} \right)} \qquad 2.17$$

An overall mass balance of the filtration gives the relation between $m_{cake,dry}$ and V_f, where $V_{cake,wet}$ is the wet volume of filter cake and c is concentration (equation 2.18).

$$m_{cake,dry} = c \left(V_f + V_{cake,wet} \right) \approx cV_f \qquad 2.18$$

A combination of equations 2.17 and 2.18 express the integrated Darcy´s Law (equation 2.19).

$$\frac{t_f}{V_f} = \frac{\eta \, \alpha_{av} \, c}{2 \, \Delta p_H \, A^2} V_f \qquad 2.19$$

In these equations α_{av} is the average specific filter cake resistance (for polymer filter cakes are highly compressible), derived from the polymer cake height and the filtration time t_f. By definition α_{av} is a limiting factor for the filtration kinetics of biopolymers. Since α_{av} is a function of the polymer-polymer interaction and the polymer-solvent interaction, it can be applied for the interpretation of the filtration data on molecular scale. Diagrams, containing the gradient of t_f/V_f versus the total filtrate volume V_f present the specific cake resistance α_{av}.

Filter cake compressibility is an important parameter to describe filtration behavior. The forces influencing the compressible biopolymers result in a more dense system with decreased cake voidage, indicating the increased specific filter cake resistance. For an incompressible cake, the cake thickness is directly proportional to the filtrate volume and inversely proportional to the filter area. However, most biological-material cakes are compressible. The compressibility of a cake increases the pressure drop across the cake, thereby strengthening the filter cake resistance. This relationship is represented by the equation 2.20:

$$\alpha = \alpha'(\Delta p)^s \qquad\qquad 2.20$$

where s is the compressibility of the cake, which can vary from 0 for an incompressible cake to 1 for a compressible cake. α' is a constant that relates to the size and shape of the biological particles forming the cake. The compressibility of the cake does change the expression of the cake resistance (García et al. 1999).

If the cake is incompressible, the plot of t_f/V_f versus V_f should give a straight line. For biotechnological products, the slopes are increasing with the volume filtered, indicating that α is increasing. However, α can increase if the cake is compressed. Cakes are typically compressible when cells and other biological materials are filtered. This reality hinders the filtration ability of the dispersions.

The other important processing parameter is concentration factor (f) characterizing the extent of the concentration. The concentration factor is defined as follows (Lewis 1996):

$$Concentration\ factor\ (f) = V_a/V_k \qquad\qquad 2.21$$

where V_a is feed volume and V_k is final concentrate volume.

Permeate rate or flux is another parameter and usually expressed in terms of volume per unit time per unit area (L/m^2h). This expression allows comparison of different membrane configurations of different surface areas (Lewis 1996).

2.4.2 Electrofiltration

Filtration combined with an electric field leads to electrophoretic migration of the dissolved or suspended substances, but also to electroosmotic effects, both on membranes and filter cakes (Moulik 1971).

Electrofiltration is highly innovative, state of the art technique for separation, respectively concentration of colloidal substances as biopolymers. The objectives of electrofiltration are to reduce the formation of filter cake and to improve the filtration kinetics of products which are difficult to filtrate. The principle of electrofiltration is based on overlaying an electric field on a standard dead-end filtration. This field acts parallel to the flow direction of the filtrate. According to the zeta potential of the substances, the direction varies towards anode and cathode. The electric field induces an electrophoretic flux of charged biopolymers towards the oppositely charged electrode. When the electrophoretic force, overruns the hydrodynamic resistance force, the charged particles migrate from the filter medium. Cake formation occurs on the membrane, on the side of the electrode with opposite to the product charge. This reduces significantly the thickness of the filter cake on the membrane, located next to the electrode with the same polarity as the product. As a result the filtration time in comparison to standard filtration is significantly reduced.

The electric field induces an electrophoretic flux of charged biopolymers towards the oppositely charged electrode. For example, in the case of a negatively charged biopolymer, the filter cake formation on the membrane next to the cathode is reduced. Only a thin surface layer is formed on this filter medium allowing permeation of almost all the filtrate through this membrane. Schematic description of electrofiltration for a negatively charged biopolymer is presented on Figure 2-12.

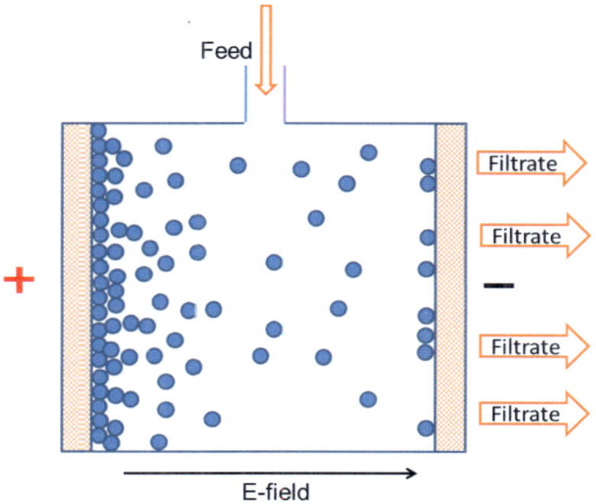

Figure 2-12: Schematic description of electrofiltration

Yukawa et al. (1976) modified Darcy's law, describing filter cake formation, with electrophoretic and electro-osmotic effects, presented with equation 2.22. This formula is valid for the filter cake free side.

$$\frac{dV_f}{dt_f} = \frac{(\Delta p_H + \Delta p_E)A^2}{\eta\,\alpha_{av}\dfrac{E_{crit} - E}{E_{crit}}m_{cake,dry}} \qquad\qquad 2.22$$

In equation 2.22, which represents adapted version of Darcy's law (equation 2.19), Δp_E is the electro-osmotic pressure and E_{crit} is the critical electric field strength, describing the equilibrium state between electrophoretic force and the hydrodynamic resistance force. This equation can be integrated by assuming that electroosmotic pressure and the critical electric field strength remains constant resulting in equation 2.23:

$$\frac{t_f}{V_f} = \frac{\eta\, \alpha_{av}\, c \left(\frac{E_{crit} - E}{E_{crit}}\right)}{2(\Delta p_H + p_E)\, A^2} V_f \qquad 2.23$$

t_f/V_f versus V_f describes the electrofiltration on the side with cake free membrane. For negatively charged products like xanthan, hyaluronic acid (HA) and poly(3-hydroxybutyrate) (PHB), t_f/V_f versus V_f describes the filtration on cathode side membrane, while for positively charged, like chitosan - on anode side membrane. Therefore the t_f/V_f versus V_f plot, presented on Figure 2-13, describes the electrofiltration on the cathode side membrane and for better understanding and clear interpretation of the processes within this system is divided into three parts.

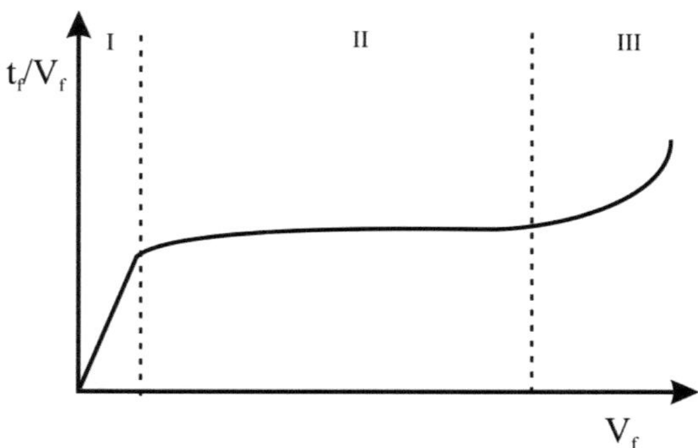

Figure 2-13: Schematic t_f/V_f versus V_f diagram of an electrofiltration experiment

<u>Section I:</u>

In the beginning of the experiment the filtrate flux is high, and no polysaccharides block the filter membranes. As a result the hydrodynamic resistance force is bigger than the electrophoretic force and thin filter film is formed on both membranes. A gradient can be observed in the first section of the t_f/V_f versus V_f plot, designating the specific filter cake resistance. Compared to conventional filtration experiment, the gradient of pressure electrofiltration experiment is significantly lower. Hence, even at this early stage of the filtration the electrical field has a positive influence on the filtration kinetics.

Section II:

Biopolymer cake formation on the membranes causes a decrease in the filtrate flux and further in the hydrodynamic resistance force. On the cathode side resistance force and electrophoretic force act in opposite directions, while on the anode side – in the same. At a certain point the resistance force acting on the biopolymer is in equilibrium with the electrophoretic force, indicating that the electric field strength is nearly equal to the critical electric field strength. After this point no accumulation occurs on the cathode-side membrane. This phenomenon can be observed in section II examining the horizontal gradient.

Section III:

The third section is characterized with an increase of biopolymer concentration in the filter chamber and the deposition of bigger biopolymer amounts on the cathode side membrane. The filter cake on the anode side reaches finally the filter cake on the cathode side and the filtrate flux reduces. That is demonstrated by the increasing gradient on the plot.

Although with different designations and slightly modified technology, the process of electrofiltration is described in the works of many scientists. In the presented work, the name "electrofiltration" designates the pressure induced dead-end electrofiltration.

Managold (1937) was first that investigated the process consisted of conventional filtration and electrophoresis. Henry (1977) then gave a fundamental analysis of cross-flow electrofiltration process applied on a kaolin clay suspension and an oil in water chemically stabilized emulsion. Additionally to biopolymer suspensions many research groups investigated dead-end electrofiltration of mineral suspensions (Moulik 1971; Yukawa et al. 1976; Mullon et al. 1985; Iwata et al. 1991; Weber and Stahl 2002; Larue and Vorobiev 2004; Saveyn et al. 2005). Recently, Hakimhashem et al. (2010) applied the dead-end electrofiltration for the removal of ecotoxix micropollutants such as pesticides and endocrine disrupters in waste water streams. Applying an electric field during the filtration process improved the filtrate flow rate considerably. An electric field of 5V/cm increased the filtrate flow rate 30 fold at 200 kPa.

In some applications researchers investigated the use of cross-flow electrofiltration of biopolymers instead of dead-end filtration technique. Yukawa et al. (1983) tested the electro-cross-flow filtration of protein solutions for the improvement of dewatering. Relations between electro-ultrafiltration flux and electric field strength were reported. The same procedure but presented as cross-flow electrofiltration was used by Brors (1992) to concentrate xanthan biosuspension. Park (2005) investigated the improvement of dead-end filtration during cross-flow electro-microfiltration of proteins. A process strategy was set up to improve recovery efficiency of proteins by dead-end filtration. The resistance in the membrane process in the presence of an electric field was reduced over 200% in comparison with the membrane process used in the absence of an electric field. In order to reduce problems with fouling, Enevoldsen et al. (2007) introduced electro-cross-ultrafiltration to concentrate industrial enzyme solutions. They reported that the flux improved 3-7 times for enzymes at an electric field of 1600 V/m compared with conventiona ultrafiltration. The effect of applying an external electric field during cross-flow microfiltration of lactoferrin and whey protein solutions was investigated by Brisson et al. (2007). The results showed the selectivity enhancements and potential of electrically enhanced membrane filtration as regards lactoferrin fractionation.

However compared to cross-flow filtration, a dead-end filtration technique exhibits not only increased permeate flow but also guarantees reduced shear forces which qualifies it as particularly mild technique for separation of biopolymers like proteins. This makes dead-end filtration preferable where cross-flow filtration can lead the denaturation of the proteins which are usually sensitive and labile (Kanani 2008). Furthermore, much higher dry matter concentration in the retentate can be obtained by dead-end filtration technique.

Experiments with dead-end ultrafiltration of protein bovine serum albumin (BSA) succeeded in reducing the membrane fouling (Iritani et al. 1992). They reported the effect of some process variables such as the solution concentration, the pH of solutions and the applied filtration pressure on the performance of dead-end electro-ultrafiltration. Bargeman et al. (2002) isolated positively charged peptides with antimicrobial activity from an alpha(s2)-casein hydrolysate using batch-wise electro-membrane filtration. Hofmann et al. (2001) presented one-side electro-pressure Nutsch filtration for the concentration of xanthan dispersions. This system evolved and improved into next generation of electrofiltration devices – two-sided electrofiltration. Changes in the process stream due to reactions at the electrodes, together with high energy requirements and considerable heat generation are the main problems associated with electrofiltration (Bowen and Ahmad 1997; Weber and Stahl 2002). In order to overcome these problems, an improved electrofiltration system by using a flushing liquid was designed (Hofmann and Posten 2003) and this innovative electrofiltration system, which also hinders the biopolymer contact with electrodes, was patented (Patent WO/2002/051874). The applicability of the described system was previously verified by the design of a pilot-scale set-up of the process (Hofmann et al. 2006).

Additionally to concentration of biopolymers, fractionation of biopolymers was tested. The fractionation of lysozym and bovine serum albumin was tested by means of electrofiltration. It was notified that the selectivity of binary separation can be increased considerably by this method (Käppler and Posten 2007). Yang et al. (2004) reported that the activity of different enzymes was reduced over electric field strength of 10 kV/cm.

Electrofiltration was applied moreover for the separation of intact microbial cells from suspensions. Ghrisan et al. (2005) tested electrofiltration on yeast suspensions and discovered that the protein content of the cultivation medium strongly influences the filtration kinetics.

3 Materials and methods

3.1 Biopolymers

The model biopolymers – PHB, chitosan, and hyaluronic acid were used for the investigation of purification and concentration of the product categories, which have not been yet tested by means of electrofiltration. These biopolymers were selected based on their commercial and technical relevance.

Chitosan and fed- batch processed PHB dispersions were produced in cooperation with SIAB (Saxon Institute of Applied Biotechnology, Leipzig) within the scope of BMBF (Federal Ministry of Education and Research) project, assigned by project number 31P556831P. PHB was produced by the bacteria *Ralstonia eutropha* in both batch and fed-batch processes. Examination of the major implications by application of different process strategies was proposed for the characterization of operational and economical future trends of PHB production. Following cell harvesting and cell disruption steps after cultivation of the microorganism, PHB dispersion was conducted to electrofiltration system.

The second investigated product, chitosan, was produced by the both fungi species *Mucor rouxii* and *Absidia coerulea* in a semi-continuous process and then transferred to the electrofiltration pressure vessel with a concentration of 2 g/L. In addition to fungal chitosan, the dispersions of crab-derived chitosan were applied to the electrofiltration system in order to determine the effectiveness of different sources upon purification process. Crab-derived chitosan was purchased from Sigma-Aldrich (Seelze, Germany) and dispersed in 1% (v/v) acetic acid with a concentration of 2 g/L for the electrofiltration experiments. In Table 3-1, cultivated microorganisms used in the production of PHB and chitosan were summarized according to origin, strain and process strategy. All microorganisms used in this work, were obtained from DSMZ (German Collection of Microorganisms and Cell Cultures, Braunschweig, Germany).

Table 3-1: Produced biopolymers with their specifications

Product	Origin	Strain	Process strategy
PHB	*Ralstonia eutropha*	DSM 428	Batch
	Ralstonia eutropha	DSM 4058	Fed-batch
Chitosan	*Mucor rouxii*	DSM 1191	Semi-continuous
	Absidia coerulea	DSM 3018	Semi-continuous

The third investigated product, hyaluronic acid, was purchased from Sigma-Aldrich (Seelze, Germany) in powder form produced by the bacteria *Streptococcus equi*. Experiments were conducted with a dispersion of 2 g/L hyaluronic acid. The applicability and relevance of electrofiltration was further confirmed by a set of analytical tools allowing the determination of product modification and structural changes of these biopolymers during purification.

For the optimization part of the experiments, xanthan, dextran and FITC-dextran which were purchased from Sigma-Aldrich (Seelze, Germany) were utilized. In Table 3-2, the purchased biopolymers that have been used in this work were summarized with their specifications in terms of origin and product number.

Table 3-2: Purchased biopolymers with their specifications

Product	Origin	Product number
Chitosan	Crabs	50494
Hyaluronic acid	Rooster combs	H5388
	Streptococcus equi	53747
Xanthan	*Xanthomonas campestris*	G1253
Dextran	*Leuconostoc mesenteroides*	D5501
FITC-dextran	*Leuconostoc mesenteroides*	FD20

Based on the previous work (Hofmann 2006) and due to its advantageous properties, xanthan with a concentration of 2 g/L was used as a model product in order to investigate the ion transfer phenomena through the membranes and the determination of effective voltage in the filter chamber. A combined dispersion of dextran and fluorescein isothiocyanate (FITC) labeled dextran with a concentration of 5 g/L was used for the optimization experiments related to the determination of concentration gradient in the filter chamber. The chemical structure of FITC-dextran is presented in Figure 3-1. The extent of labeling was 0.003-0.020 mol FITC per mol glucose. FITC-dextran as a labeled biopolymer enabled the visualization of the concentration gradient by the fluorescence sensors which were integrated into the filter chamber.

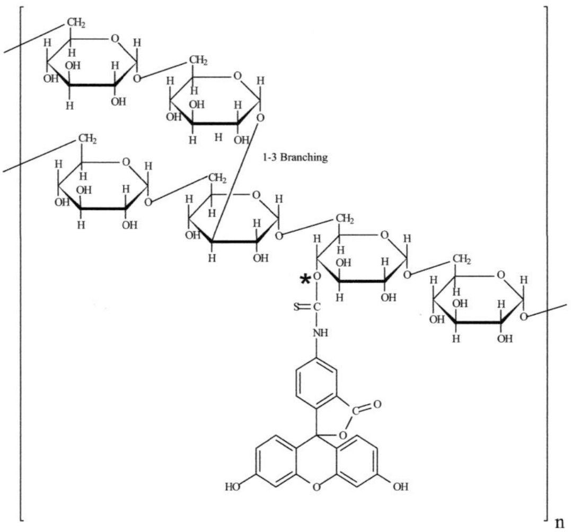

* The site of attachment of FITC is assumed to be randomly associated with any free hydroxyl group

Figure 3-1: Chemical structure of FITC-dextran (Sigma-Aldrich)

3.2 Cultivation procedure

3.2.1 Growth of PHB containing cells

Batch process

Batch cultivation of the bacterium *Ralstonia eutropha* strain ATCC 17699 (DSM 428) were carried out in a 15 L bioreactor (B. Braun Biotech, Melsungen, Germany) with a working volume of 6 L by using a synthetic medium which consisted of 20 g/L sodium gluconate as carbon source, 9.0 g/L $Na_2HPO_4 \cdot 10H_2O$, 1.5 g/L KH_2PO_4, 1.0 g/L NH_4Cl, 0.2 g/L $MgSO_4 \cdot 7H_2O$, 17.6 mg/L $CaCl_2 \cdot 2H_2O$, 1.2 mg/L ammonium ferric citrate ($C_6H_{11}FeNO_7$), and 0.1 mL/L of 100-fold prepared SL6 trace element solution (see Table 3-3). Sodium gluconate, $MgSO_4 \cdot 7H_2O$, $CaCl_2 \cdot 2H_2O$ and ammonium ferric citrate solutions were autoclaved separately in order to avoid undesirable chemical reactions of medium components and then transferred to the sterile medium in bioreactor.

Inoculum was prepared in two steps. The first stage inoculum was grown in 50 mL nutrient broth medium (5 g/L peptone, 3 g/L meat extract) at pH 7 and incubated for 15 h by gyratory shaking (120 rpm) at 30 °C. Afterwards, the culture was transferred into 150 mL mineral salt medium (see Table 3-3) and incubated for 24 h at same conditions (the second stage inoculum). After sterilization of the medium, the bioreactor was inoculated with 400 mL of the final inoculum. During the course of cultivation, temperature and pH was kept at 30°C and 7.0, respectively. The other conditions used were agitation at 400 rpm, aeration at 1.67 Nl/min. A silicone based antifoam agent with a volume of 4 mL (Antifoam 204, Sigma-Aldrich) was added to control foaming in the bioreactor. The fermentation was monitored online by measuring partial pressure of dissolved oxygen (pO_2) and carbon dioxide (pCO_2) with the beginning of the exponential growth.

The optical density at 550 nm was recorded offline. Based on the fact that PHB is accumulated intracellular by *Ralstonia eutropha* in the stationary phase (Hänggi 1990), the cells were harvested in the stationary phase, centrifuged and washed twice. Considering that *R. eutropha* produces PHB under limitation of essential nutrients (Rebah et. al. 2004; Johnson et al. 2010), nitrogen uptake was monitored photometrically through measuring ammonia concentration of supernatant cultivation probes. The used ammonia assay (NH_4^+ Spectroquant®, Merck) was modified for the available sample volume.

Table 3-3: Medium components of batch cultivation for production of PHB containing cells

Media	Components	Concentration	Supplier	Literature
Synthetic medium	Sodium gluconate	20.0 g/L	Merck	Schlegel et al. 1961
	Di-sodium hydrogen phosphate dodecahydrate, p.a.	9.0 g/L	AppliChem	
	Potassium dihydrogen phosphate, ≥ 99%, p.a.	1.5 g/L	Roth	
	Ammonium chloride, ≥ 99.5%	1.0 g/L	Roth	
	Magnesium sulfate heptahydrate, p.a.	0.2 g/L	AppliChem	
	Calcium chloride dihydrate, p.a.	0.0176 g/L	AppliChem	
	Ammonium ferric citrate, 18% Fe	0.0012 g/L	Roth	
	SL6 (100 fold)	0.1 mL/L		
Nutrient medium	Soya peptone	5 g/L	Roth	Sambrook et al. 1989
	Meat extract	3 g/L	Roth	
SL6 trace elements	Zinc sulfate heptahydrate	10 mg/L	Merck	Pfennig 1974
	Manganese(II)-chloride tetrahydrate	3 mg/L	Merck	
	Boric acid	30 mg/L	Roth	
	Sodium molybdate dihydrate	3 mg/L	Merck	
	Cobalt(II)-chloride hexahydrate	20 mg/L	AppliChem	
	Copper(II)-chloride dihydrate	1 mg/L	Fluka	
	Nickel(II)-chloride hexahydrate	2 mg/L	Roth	

Fed batch process

Fed-batch cultivation of *Ralstonia eutropha* strain JMP 134 (DSMZ 4058) with a PHB content of 50% was carried out in a mineral medium (pH 7.0) at 35°C as described in Mothes et al. (2007). The cultivation was performed in a 5 L bioreactor (B. Braun Biotech, Melsungen, Germany) with the medium contained 0.95 g/L $(NH_4)_2SO_4$, 5.47 mg/L $CaCl_2 \cdot 6H_2O$, 0.68 g/L KH_2PO_4, 0.87 g/L K_2HPO_4 and 1 mL/L sterile stock solution of trace elements which consisted of 71.20 g/L $MgSO_4 \cdot 7H_2O$, 0.78 g/L $CuSO_4 \cdot 5H_2O$, 0.44 g/L $ZnCl_2 \cdot 7H_2O$, 0.81 g/L $MnSO_4 \cdot 4H_2O$, 0.25 g/L $Na_2MoO_4 \cdot 2H_2O$, and 4.98 g/L $FeSO_4 \cdot 7H_2O$ (see Table 3-4).

Table 3-4: Medium components of fed-batch cultivation for production of PHB containing cells

Media	Components	Concentration	Supplier	Literature
Mineral medium	Glycerol	10 g/L	EOP Biodi.	Mothes et al. 2007
	Ammonium sulfate, p.a	0.95 g/L	Roth	
	Calcium chloride hexahydrate, p.a	5.47 mg/L	AppliChem	
	Potassium dihydrogen phosphate, ≥ 99%, p.a.	0.68 g/L	Roth	
	Dipotassium hydrogen phosphate	0.87 g/L	AppliChem	
Trace elements	Magnesium sulfate heptahydrate	71.20 g/L	AppliChem	Mothes et al. 2007
	Copper sulphate penta hydrate	0.78 g/L	AppliChem	
	Zinc chloride heptahydrate	0.44 g/L	AppliChem	
	Manganese(II)-sulphate tetrahydrate	0.81 g/L	Merck	
	Sodium molybdate d hydrate	0.25 g/L	Merck	
	Ferrous(II)- sulphate heptahydrate	4.98 g/L	Merck	

As carbon source 10 g/L *Glycerin* (EOP Biodiesel GmbH, Falkenhagen) composed of 85% glycerol, 9.2% water, 0.03% methanol, 0.8% K_2SO_4 was used at a pH value of 4.1. Ammonium sulphate, calcium chloride and glycerol were autoclaved in the bioreactor. Then, the separately sterilized trace element solution and phosphate were added and subsequently the inoculum culture, which was incubated in 500 mL shaking flasks in the defined medium (Table 3-4), was transferred into the bioreactor. To reach a higher cell density, the trace element solution and phosphate were added up to 18-fold and 2-fold, respectively. During the growth phase, 12.5% NH_4OH was fed continuously for pH correction. In order to initiate N-limited PHB synthesis, NH_4OH was replaced with 2 N KOH. Glycerol was fed continuously to maintain a consumption rate.

3.2.2 Growth of chitosan containing cells

Mucor rouxii (DSM 1191) and *Absidia coerulea* (DSM 3018) were maintained on 3.9% potato dextrose agar (PDA) slants at 4°C. Biomass for inoculating the submerged culture was grown on PDA slants at 30°C for 96 h. Suspensions of spores were prepared by re-suspending slants with sterile, distilled water. The nutrient broth used in the experiments consisted of 40 g glucose, 10 g peptone, 1 g yeast extract, 4.8 g $(NH_4)_2SO_4$, 0.5 g $MgSO_4 \cdot 7H_2O$, 0.2 g $CaCl_2$, 50 mL tap water and 950 mL distilled water (see Table 3-5).

Cultivation was conducted as described in Trutnau et al. (2009) in a 30 L stirred tank reactor (B. Braun Biotech, Melsungen, Germany) with two pitched blade turbines. Two different inoculum types were added to a sterilized medium to a total volume of 20 L. Inoculums contained either a suspension of 1×10^8 spores/L, or three submerged subcultures, prepared by incubating 200 mL of broth with a suspension of 5×10^7

spores/L in a 1 L flask on an incubation shaker (Infors, Switzerland) operating at 150 rpm, 28°C for 24 h. Cultivating conditions of 28°C, pH 4.5, 700 rpm stirring rate and 2 vvm aeration rate (volume air/volume medium/min) were maintained during the process.

Table 3-5: Medium components of semi-continuous cultivation for production of chitosan containing cells

Medium	Components	Concentration	Supplier	Literature
Nutrient broth	Glucose, 99.5%	40 g/L	Roth	Trutnau et al. 2009
	Soya peptone	10 g/L	Roth	
	Yeast extract	1 g/L	Roth	
	Ammonium sulfate, p.a	4.8 g/L	Roth	
	Magnesium sulfate heptahydrate, p.a.	0.5 g/L	AppliChem	
	Calcium chloride	0.2 g/L	AppliChem	

Cultivations were performed either as single batch or semi-continuous (repeated batch) processes. For the latter, process times were 24 hours for subculture-inoculated batches, or 40 hours for spore-inoculated batches. The bioreactor contents were then removed for further processing, with a residue of 1000-2000 mL as an inoculum for the subsequent batch with fresh nutrient broth (20 L). This procedure could be repeated between two and seven times. Samples for biomass growth determination were taken first after 14 h of incubation for spore inoculated cultures and after one hour of incubation for submerged subculture inoculated cultures. In both cases, the biomass of the cultures (cell dry weight) was determined as follows: three replicate samples of 30 mL were taken every hour from the bioreactor, filtered by vacuum filtration on a pre-weighed filter and washed twice with distilled water. The samples were then dried for 24 h at 105°C and weighed. The concentration of reducing sugars in the samples was also determined, using a method that has been previously described in detail (Miller 1959).

3.3 Harvesting methods

3.3.1 Release of PHB by cell disruption

PHB granules are covered with a surface layer consisting of proteins and phospholipids (Jendrossek and Handrick 2002). The surface layer of PHB granules is essential for the further purification by electrofiltration. The alternative cell disruption methods by using chemicals such as detergents and solvents, can lead to denaturation of the surface layer of PHB granules. A precise cell disruption by French press can prevent the denaturation effect (Jendrossek and Handrick 2002). Following the described circumstances, the intracellular PHB granules were released by using a French press (Electron Corporation, Needham, USA) with a cell volume of 35 mL. The harvested cells were resuspended in an adequate amount of 50 mM Tris/HCl buffer (Roth, Germany) at pH 9. The cells were disrupted by two passages through the cold French pressure cell at 2000 psi. Verification of cell disruption after each passage was done by determination of the total protein concentration by BCA (bicinchoninic acid) assay.

3.3.2 Solvent extraction of chitosan

Mycelia from every culture were harvested by vacuum filtration and washed with distilled water until the filtrate was clear. The biomass was freeze-dried and divided into aliquots (10-25 g). To each aliquot, 1 L 2 N NaOH was added to extract proteins and other unwanted cellular components, stirred for 3 h at 70°C and the mixtures were centrifuged at 6000 rpm for 10 min to sediment the alkali insoluble materials (AIM), which were washed and centrifuged twice with distilled water. The deproteinated samples were then demineralised as follows: freeze dried AIM was added to HCl solution (2 N; 1 L), stirred for 3 h at 70°C, then the resulting mixtures were centrifuged and washed, as described above.

The freeze-dried residues were suspended in 50% NaOH and stirred for 2 h at 100-130°C to deacetylate the chitin residues and to recover all the chitosan from approximately 25% of the glucosamineous components (chitin + chitosan) that are bound to the cell wall (Bartnicki-Garcia and Nickerson 1962). The deacetylation step was followed by acid extraction in 1 L 10% acetic acid for 6 h at 70°C (Rane and Hoover 1993). These conditions facilitate the drop out of the structural components (chitin, glucan) with chitosan remaining soluble.

The solution was divided in two parts: one fraction for electrofiltration and the other fraction for precipitation. NaOH (32%) was then added to the supernatant to increase the pH to 8-9 in order to precipitate chitosan, which was subsequently centrifuged and washed with 200 mL distilled water, twice with 200 mL ethanol/distilled water (20:80), twice with 200 mL acetone/distilled water (20:80) and freeze dried. These purification steps were required to ensure that the chitosan obtained corresponded to the quality criterion chosen.

3.4 Electrofiltration equipment design

3.4.1 Components

Electrofiltration system consists of three main components which are connected to each other during the experiments. The filter chamber in between is the heart of the system and framed on both sides by flushing chambers (see Figure 3-2).

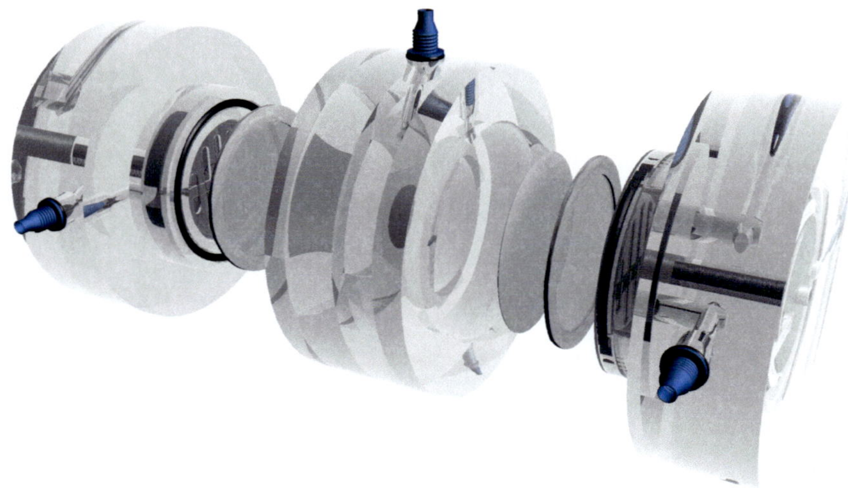

Figure 3-2: The major electrofiltration cell

The electrodes generating electric field to the system are placed inside the flushing chambers. A liquid (buffer solution) with conductivity equal to the biopolymer dispersion's one, flushes the electrodes. The function of the buffer solution is to keep pH value constant and rinse out electrolysis products, formed at the electrodes. Moreover, the flushing liquid acts like a heat exchanger, dissipating the heat, generated by the electric current, out of the filter chamber and preventing overheating of the filter cake. The flushing chamber system avoids direct contact of biopolymer with electrodes.

Chambers were produced by using the material polymethylmethacrylate due to its elasticity, transparence and resistance to chemicals. Between filter chamber and each of the flushing chambers, there are several elements that have to be considered. The filter chamber was bordered on both sides by microfiltration membranes with a distance of 10 mm between them. Next to the membranes, supporting meshes were used to avoid breakage of the membranes. The circuit of the supporting meshes was enclosed by using teflon tape to prevent leaks. Next to the supporting mesh, an electrode lattice which supports the flushing system was used. The distance between the electrodes was 20 mm indicating that the 80 V of applied voltage corresponded to an electric field strength of 4 V/mm.

Electrode materials were IrO_2 coated titanium from company Metakem (Usingen, Germany) and stainless steel (V4A) for anode and cathode, respectively. These electrode materials were selected owing to their low formation of electrolysis products. The screws that holds the electrodes were made of stainless steel (without any brass in the composition) to prevent possible electrochemical reactions. The configuration and position of membranes, supporting meshes, lattice and electrodes were presented in Figure 3-3. The hydrophilic polyethersulfone (PES) membranes with a pore size of 0.1 µm (Supor®100) from company Pall (East Hill, NY, USA) were used generally for all filtration/electrofiltration experiments. PES membrane is a highly porous asymmetric membrane that delivers high flow rates extending the capacity of the membrane.

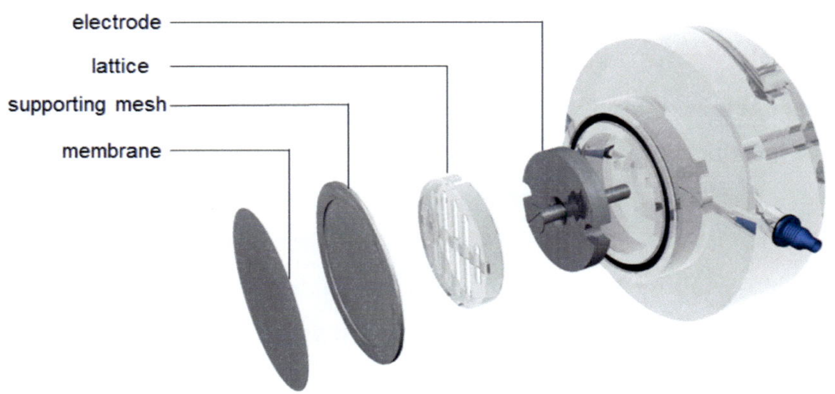

Figure 3-3: The configuration and position of electrofiltration elements between chambers

The supporting mesh was purchased from company SEFAR with a product number 04-1100-SK 028. In addition to the PES membrane and the defined supporting mesh, some other materials with a pore size of 0.1 µm (see Table 3-6) were tested in order to optimize the overall energy consumption which can be affected by the type of the materials.

Table 3-6: Membrane properties

Supplier	Material
Pall	Polyethersulfone
Millipore	Polycarbonate
Millipore	Mixed cellulose esters
Millipore	PVDF

3.4.2 Experimental implementation

A laboratory-scale apparatus was used for the investigations of electrofiltration. The filter chamber was illustrated in Figure 3-4 schematically with the migration of a positively charged molecule. The introduction of electrofiltration in biotechnological processes is particularly attractive for the purification of biopolymers, which are difficult to filtrate, and usually charged by the presence of amino and carboxyl groups in their molecules.

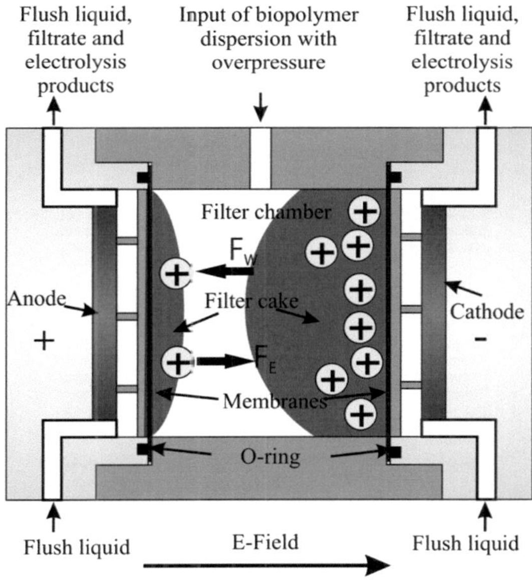

Figure 3-4: Electrofiltration chamber with the behavior of positively charged molecules

When the electrophoretic force (F_E) overruns the hydrodynamic resistance force (F_W) (see section 2.3) the charged particles migrate from the filter medium, reducing significantly the thickness of the filter cake formed. For a positively charged molecule such as chitosan, filter cake formation is reduced on the anode side and only a thin film occurs, allowing higher filtrate fluxes, due to the opposite direction of the two forces acting on positively charged molecules. However, on the cathode side, due to the same direction of the forces, the filter cake grows continuously until the end of filtration (see Figure 3-4). Therefore, filtration flux through the anode side membrane was not hindered by the blocking of the membranes. The system periphery consists of a filter chamber, a voltage generator, a pressure vessel which can be cooled by a mantel to prevent denaturation of the product, a peristaltic pump to circulate the flushing liquid, a reservoir to collect the filtrate and a balance connected to the computer to measure the filtrate mass (see Figure 3-5).

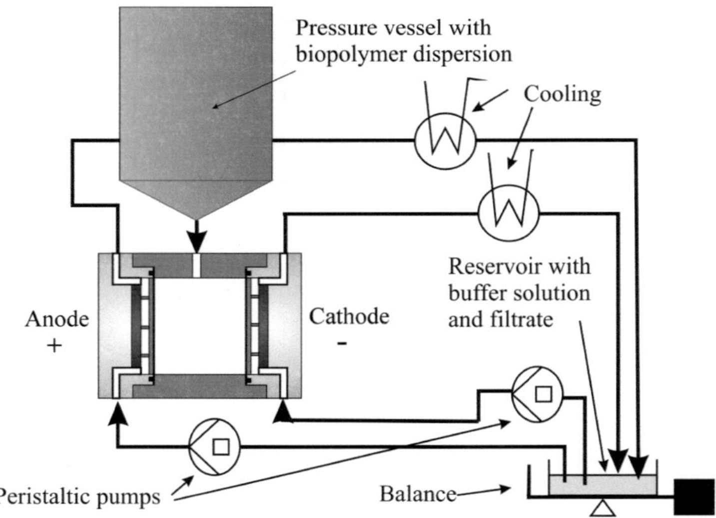

Figure 3-5: Electrofiltration system

Before starting the experiment, membranes and supporting meshes were wetted with the buffer solution that will be used in the experiment in order to increase the permeate flux. After preparation of filter media, chambers were fixed to each other by rotating the valve. The tube transporting biopolymer dispersion from the pressure vessel into the filter chamber was filled with the dispersion. Before conducting pressure to the system, the filter chamber was also filled with the biopolymer dispersion and then the buffer solution, which had a start volume of 500 mL in the reservoir, was circulated through the silicone tubes. At the beginning of the experiment, tare weighed so that the measured mass revealed the filtrated dispersion. Biopolymer dispersion was transported under overpressure from the pressure vessel into the filter chamber. For the proper process conduction online control of filtrate mass and pH value, together with the conductivity levels was required. The control was operated by the multiple versions of software programs Lab-View and BioProCon. The filtrate was collected in the reservoir together with the circulated buffer solution. The flushing buffer was pumped continuously by peristaltic pumps at 200 rpm (Watson Marlow 323, Rommerskirchen, Germany). By means of peristaltic pumps, the buffer solution was directed from bottom towards the top of the filter chamber to avoid the gas bubbles which can affect the process negatively. The temperature of the solution in reservoir was kept at 15°C by a cooler rig. Information about filtration kinetics was provided by measurement and evaluation of filtrate mass which was influenced by the variation of applied voltages and pressures.

During investigations of electrofiltration, different buffer solutions were used according to the product properties. In Table 3-7, the used buffer solutions with their preparation concentration and the conductivity values of the product dispersions were presented for all the experiments. For the conductivity of chitosan, the first and second values in the table demonstrate the crab-derived and fungi-derived conductivities, respectively.

Table 3-7: Used buffer solutions for different biopolymers

Product	Buffer solution	Conductivity [µS/cm]
PHB	0.033 M Tris/H_3PO_4	~ 240
Chitosan	0.1 M citric acid monohydrate and 0.2 M Na_2HPO_4	~ 600 and ~1500
Hyaluronic acid	1/15 M Phosphat Buffer	~ 300
Xanthan	1/15 M Phosphat Buffer	~ 300
FITC-dextran	1/15 M Phosphat Buffer	~ 690

The conductivity and pH of the flushing solutions were equalized generally with those of the product dispersions. The pressure vessel was kept at room temperature for all of the biopolymer dispersions except chitosan dispersions. Due to the degradation and denaturation of the chitosan molecules in the absence of cooling, the dispersion in the pressure vessel was cooled during the experiments.

3.4.3 Electrofiltration cell constructed with voltage sensors

The conventional electrofiltration cell which was illustrated in Figure 3-2 was used for the investigation of the purification of the chosen biopolymers and the experiments related to the migration of the ions. On the other side, investigation of *in-situ* behavior of electrofiltration plays an important role for the further improvement of the system. Based on this, a novel filter chamber was designed and constructed as illustrated in Figure 3-6. The objective of the novel design was to determine the voltage drops in the system in order to acquire information about development of the internal structure of the separation process for optimization of the process parameters and reduction of the production costs. The design comprises 7 platinum voltage sensors, which of 5 placed into the filter chamber and 2 very close to the electrode surfaces.

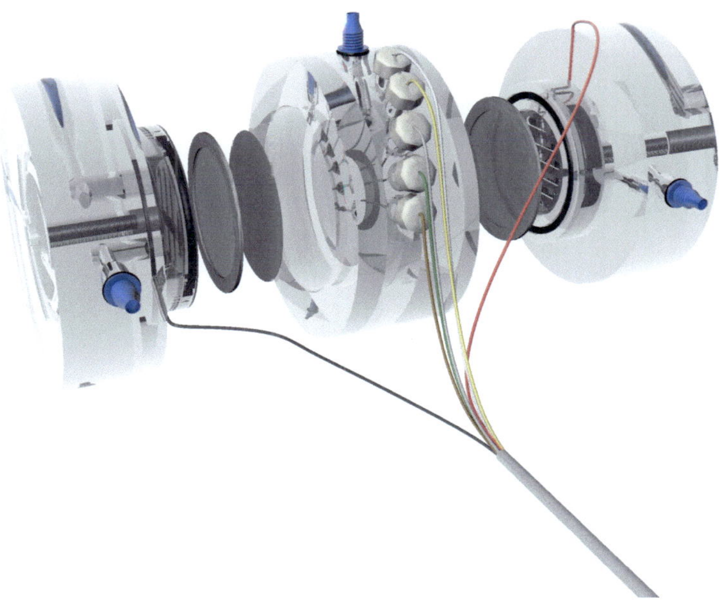

Figure 3-6: Electrofiltration cell constructed with voltage measuring electrodes

The perspective of the voltage sensors in the filter chamber was illustrated in Figure 3-7. All of the sensors were connected to the micronet and the values during the experiments were saved online. The sensors in the filter chamber were placed regularly with a distance of 2 mm. The other 2 sensors had a distance of 1 mm to the working electrode surfaces. The distance between the first and second sensors from both sides was higher due to the width of the flushing chambers. For a better evaluating, each of the voltage differences was named according to the positions in the electrofiltration cell. In Figure 3-8 the position of the 7 sensors with their distance in mm to the anode electrode was demonstrated in details.

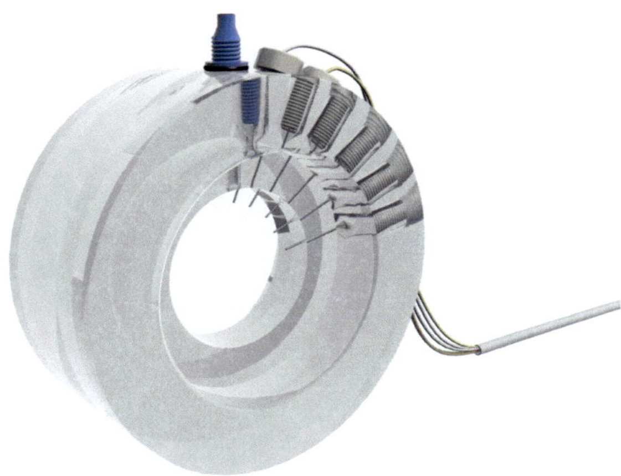

Figure 3-7: The position of the voltage measuring electrodes in the filter chamber

The numbers for each segment presents the cross sectional length of the segments in the cell. U_A and U_C, for anode and cathode respectively, were the voltage differences between the sensors close to the membrane surfaces in the filter chamber and the sensors close to the electrode surfaces.

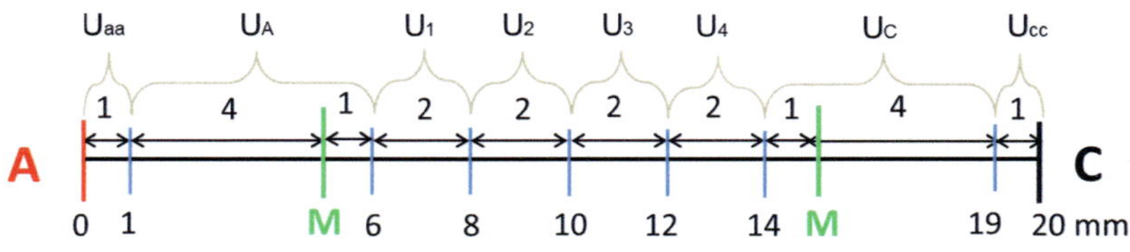

Figure 3-8: The distances between the voltage sensors through the system

The elements which stated in this segment i.e the supporting mesh, membrane and electrode lattice were the elements that affected the voltage drop on the membranes. The sum of U_1, U_2, U_3 and U_4 namely U_{eff} reveals the actual voltage in the filter chamber. A, C an M corresponds to the anode, cathode and membranes, respectively. U_{aa} and U_{cc} were estimated by subtraction of the values online measured from the applied voltage and if necessary manually.

3.4.4 Electrofiltration cell constructed with fluorescence sensors

For the further *in-situ* investigation of electrofiltration system another novel chamber was designed and constructed. The objective was the optimization of the electrofiltration process by understanding the mechanisms and phenomena occurring inside the filter chamber. Modifying apparatus design was expected to bring information about the concentration gradient in the filter chamber by application of an electric field. The illustrated electrofiltration chamber in Figure 3-9 was designed and constructed in a cooperation with the company Optoprim (Landsberg am Lech, Germany).

Fluorescence as an investigative tool was applied due to its high capability in studying the structure and monitoring of the materials. Therefore, analysis of the internal structure of the filter cake and the migration of the biopolymer molecules during electrofiltration was studied by fluorescence technique. After a series of the pre-experiments with different probe end materials and different measurement principles, carbon tips working with reflection principle were chosen. The probes fixed to the filter chamber had a diameter of 1 mm with 2x200 µm UV optic fibers. Carbon was preferred as material, due to its chemical resistance in the presence of electric field. Additionally, carbon tips had a relative smaller diameter than the other tested materials such as ceramic. The carbon tip was stabled into the chamber by M6 nylon threads. LED light source (at 460 nm) were used for the experiments allowing evaluation of emission and reflection peaks.

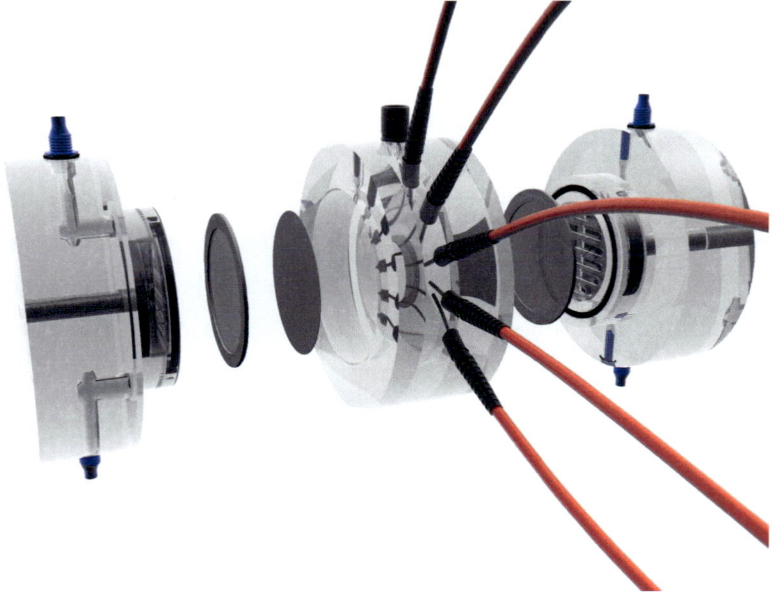

Figure 3-9: Electrofiltration cell constructed with fluorescence sensors

The perspective of the probes in the filter chamber was demonstrated in Figure 3-10. The illustrated system comprises 5 sensors with a 2 mm distance to the next one. In Figure 3-11, the distances between the tips were demonstrated in details. The tips were named starting from the cathode towards anode side.

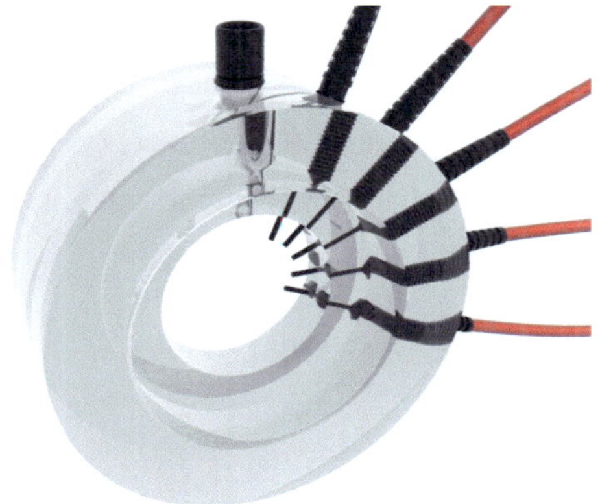

Figure 3-10: The position of the fluorescence sensors in the filter chamber

Pre-experiments were performed to determine the optimum fluorescence marker and mixing ratio. After testing several labeled biopolymers with different molecular weights and compositions, experiments were conducted with 5 g/L mixture of dextran and FITC labeled dextran dispersion.

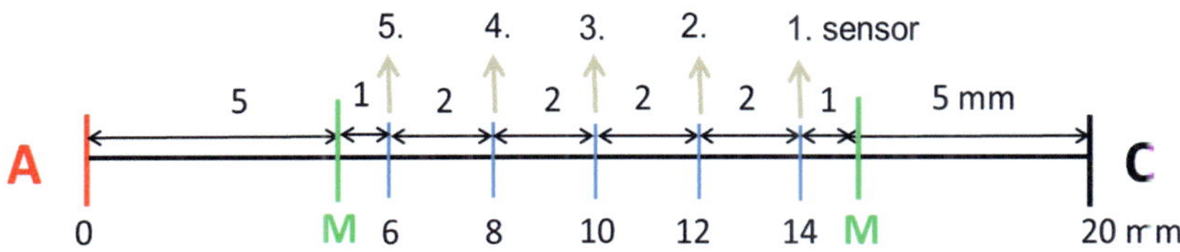

Figure 3-11: The distances between the fluorescence sensors through the system

The filter chamber was covered with no light entrance completely to prevent disturbance of the spectrum. During the electrofiltration experiment, the migration of the marked biopolymer inside the filter chamber was monitored online and the data enabled by the measurement of the fluorescence inside the filter chamber was evaluated by the software program Avasoft version 7.5. The modeling of the composition and distribution of the filter cake in the chamber was developed by the simulation program MATLAB.

3.5 Characterization methods

3.5.1 Determination of dry matter

Prior to freeze drying, both the sample and vessel masses were determined by an analytical balance (AC 210 S, Sartorius AG, Göttingen, Germany). Samples were frozen at −30°C in a freezer and then lyophilized (Alpha 1-4, Christ Gefriertrocknungsanlagen, Osterode, Germany) at −54°C and a pressure of <2 mPa. Samples were freeze-dried up to 18 h, depending on the stability of the pressure. In cases of thicker filter cakes, freeze-drying continued for 48 h. After attaching vessel covers, samples were weighed again. The water content was determined from the mass difference before and after freeze-drying. The dry matter was calculated by subtracting the mass of the vessel alone. Random controls confirmed that sample's mass remained constant after the first freeze-drying.

3.5.2 Degree of deacetylation (D_{DA})

After dissolving 100 mg chitosan by overnight stirring in 10 mL 0.1 M hydrochloric acid, 25 mL bidistilled water were added. This solution was titrated to a pH of 3.20 to 3.25 with 0.1 M sodium hydroxide using a pH electrode (Minitrode, Hamilton Company) and a pH meter (pH-Meter 706 Calimatic, Knick Elektronische Messgeräte & Co. KG, Berlin, Germany).

The D_{DA} could be estimated by the amount of sodium hydroxide used and the original amount of hydrochloric acid and calculated by using different equations. The following formula was derived from dimensional analysis (mathematical method) whereas the molecular weights M [g/mol], the sample mass m [g], the molarity c [mol/L] and volume V [L] of the NaOH and HCl solutions are used, f stands for dimensionless term (Trutnau et al. 2009).

$$D_{DA} = \frac{m_{Sample} \cdot M_{acetyl\text{-}glucosamine} \cdot f \cdot c \cdot (V_{NaOH} + V_{HCl})}{f \cdot c \cdot (V_{NaOH} + V_{HCl}) \cdot (M_{acetyl\text{-}glucosamine} + M_{glucosamine})} \cdot 100$$

3.1

$$with\ f = 331.310 \frac{V_{NaOH}^2}{[L]^2} - 26.243 \frac{V_{NaOH}}{[L]} + 0.96$$

3.5.3 Protein content

The protein content was determined for the samples of PHB after cell disruption and chitosan after electrofiltration. Based on a 96-well plate reading (Smith et al. 1985), the protein content in aqueous samples was determined by the bicinchoninic acid assay (BCA). Briefly, a mixture of 10 mL BCA Reagent A (Interchim, France) and 200 µL 4% copper sulfate solution, designated further as "BCA reagent", was used. In order to remove all unsolved residues, samples were centrifuged at 13400 rpm (approx. 12000 g) for 5 min. Subsequently, 10 µL of the sample were introduced into each well, followed by 200 µl BCA reagent. Bovine serum albumin (BSA) was used as a standard for the protein content estimation. The plate was incubated for 30 min at 37°C and read at 560 nm using a photometer (Multiskan Ascent, Thermo Fisher Scientific, Inc.). All results were derived after threefold sample determination.

3.5.4 Dynamic viscosity

Samples of 50 mg chitosan were dissolved in 25 mL 1% acetic acid and stirred for 24 h at 850 rpm (Variomag Electronicrührer MULTIPOINT HP, H+P Labortechnik, Oberschleißheim, Germany). Cloudy solutions were centrifuged for 10 min at 2000 rpm in 50 mL Falcon tubes (BHG Hermle ZK380, HERMLE Labortechnik GmbH, Wehingen, Germany). The supernatant was carefully removed and used for viscosity measurements.

The dynamic viscosity was determined in a couette rheometer (Physica MCR 301 with Messsystem CC27, Anton Paar GmbH, Graz) at shear rates of 1 to 1000 L/s, 10 s per shear rate and six points per order of magnitude. This resulted in 19 points per measurement and each solution was measured three times. Shear rates started at low values and increased during measurement. The data was collected by the software RHEOPLUS V3.10 (Anton Paar GmbH, Graz). With a measurement temperature of 25.00 °C ± 0.04°C, the viscosity was determined from the horizontal part at a shear rate of 46.4 L/s.

3.5.5 Molecular weight distribution by SEC-MALLS

Samples were dissolved in acetate buffer (0.1 M sodium acetate and 0.15 M acetic acid, buffer A) at a concentration of 1 g/L by stirring for 3h at room temperature. In order to remove all unsolved residues, samples were filtered over a GHP membrane (PALL Corporation, USA) with a pore size of 0.45 µm. The molecular weights were determined by size exclusion chromatography (SEC) combined with multi-angle laser light scattering (MALLS) and refractive index (RI) detection. The system was composed of an Agilent 1100 Series Chromatography system with a RI detector using a TSKgel G6000PWxl column (Tosoh Bioscience GmbH, Stuttgart, Germany) 7.8 mm ID and 30 cm length with a guard column and a MALLS detector (DAWN DSP, Wyatt Technology Corporation, Santa Barbara, CA, USA). Toluene (VWR International GmbH, Darmstadt, Germany) was used for the calibration of the DAWN DSP device and normalization of the MALLS detectors was performed using BSA (5 mg/mL) in buffer A. RI calibration constants were generated with anhydrous sodium chloride. Analyses were performed at room temperature using buffer A as the mobile phase at a flow rate of 0.8 mL/min. 100 µL sample volume were applied on the column and the average of three replicates was used for the evaluation.

3.5.6 FT-Raman spectroscopy

The used FT-IR spectrometer Vertex 80 with attached Raman module RAM II (Bruker Optik GmbH, Ettlingen, Germany) employed a 1064 nm NdYAG excitation laser, an interferometer and a high sensitivity near infrared Ge-detector. For Raman analysis an adequate amount of each sample was lyophilized for at least 20 h and thawed in a desiccator. The obtained powder was transferred in a small sampling pan and analysed. The applied laser power was manually adjusted for each sample individually in a range of 20-300 mW on basis of the recorded interferogram. The received spectra were evaluated by OPUS 6.5 software (Bruker Optik GmbH, Germany). The purchased PHB and chitosan (natural origin, Sigma-Aldrich) and the untreated PHB lysate samples of batch and fed-batch were used as references.

3.5.7 Inductively Coupled Plasma - Optical Emission Spectrometry (ICP-OES)

All samples were analyzed for cadmium (228.8 nm), lead (220.4 nm), zinc (213.9 nm) and copper (327.4 nm) by a ICP-OES apparatus model JY38S, 1200W (Horiba Jobin Yvon, France) equipped with a U5000AT+ Ultrasonic Nebulizer (Cetac, Omaha, Nebraska 68144 USA) using the following conditions: carrier gas-argon, pump rate of 2.5 mL/min, plasma rate of 15 L/min, nebulizer gas rate of 0.7 L/min, auxiliary gas rate of 0.55 L/min and a carrier gas rate of 0.2 L/min. For all experiments, 10 mg of the dry mass of chitosan preparations were used.

3.5.8 Zeta potential

Zeta potential of all biopolymers investigated was determined by using a Malvern Zetasizer 5000 with PCS software V1.52, Rev1 (Malvern Ltd, UK). After double flushing of the flow cell, results were generated by the average of five measurements. A Malvern standard solution was used for result confirmation.

3.5.9 Gel Permeations Chromatography (GPC)

Molecular weight measurements of each cultivated and the purchased PHB was analysed on a GPC system containing one analytical precolumn (PSS-SDV, 5 µm, I.D. 8 mm, length 50 mm) and a combination of three analytical columns (PSS-SDV, 5 µm, I.D. 8 mm, length 50 mm) for the definition of low molecule masses. The system was equipped with a Shodex RI 71 refractive index detector and 100 µL sample volume were injected by TSP AS3000 injector. The experiments were conducted in $CHCl_3$ at 23°C and a flow rate of 1 mL/min was used.

The sample concentration in $CHCl_3$ was about 5 g/L and each sample was filtrated before GPC analysis using a 4.5 µm Teflon filter (Wateman).

3.5.10 Gas Chromatography/Mass Spectrometry (GC/MS)

PHB concentration after electrofiltration was determined by GC/MS trough depolymerisation and derivatisation of PHB to (R)-3-hydroxybutyric acid methyl ester (3-HBME) according to the method described by Braunegg et al. (1978). Before derivatisation interfering water was removed by lyophilisation of 1.5 mL well mixed sample for 24 h. Subsequent each sample was dissolved in 2 ml acidified methanol (3% v/v H2SO4) and incubated at 100°C for 3.5 h. Samples were cooled down on ice and 1 mL demineralised water and 2 mL chloroform were added. The extraction of the originated 3-HBME was done by vigorously shaking each sample for 5 min. After phase separation the required volume of 3-HBME containing phase (chloroform phase) was separated and dried at 40°C under a gentle nitrogen stream. The residual methyl ester was resuspended in 1 mL methanol and an aliquot was injected for GC/MS analysis. Each sample was analysed three times and a calibration was done in combination with all passages. The obtained dry biomass (DBM) was determined gravimetrical and used for further calculation. The GC/MS-analysis was performed on Saturn 2000 (Varian) equipped with a polar, cross linked polyethylene glycol column (Stabilwax®, Restek, 30 m by 0.25 id) and a ion trap EI-MS. Helium 6.0 was used as carrier gas. The MS-identification was accomplished by four main masses (m/z 43, 74, 103 and 119) in which the mass of m/z 74 was used for peak quantification and the other three masses were set for peak qualification. No internal standard was used. The PHB generation in biomass during fed-batch cultivation was determined by gas chromatography of (R)-3-hydroxybutyric acid propyl ester (3-HBPE) produced by propanolysis of lyophilised cells (Riis and Mai 1988). The GC (model GC-14-AX SHIMADZU Corp., Kyoto, Japan) was equipped with flame ionisation detector and Permabond FFAP column (25 m x 0.25 mm; MACHEREY-NAGEL, Düren, Germany). The carrier gas was N_2 and benzoic acid was used as internal standard.

3.5.11 Ion chromatography

The determination of inorganic anions (chloride, nitrate, sulfate and phosphate) was done by using an ion chromatography system (Dionex IC Workstation, Sunnyvale, CA, USA) equipped with a LC20 chromatography enclosure, CD20 conductivity detector, GP40 gradient pump and AS40 auto sampler. The ion separation was accomplished by using a Dionex AS11 anion exchange column in combination with a Dionex AG11 guard column at NaOH gradient (8 to 35mmol). A sample volume of 10µL (full loop) was applied on the column. For data evaluation the Dionex chromatography data system Chromeleon 6.8 was used.

4 Filtration kinetics and characterization of biopolymer separation

4.1 Separation of PHB

4.1.1 Microbial production of PHB in batch and fed-batch processes

The biodegradable polyester poly-(3-hydroxybutyrate) (PHB), produced by the bacteria *Ralstonia eutropha* in batch and fed-batch processes, was purified by means of electrofiltration.

A batch cultivation of *R. eutropha* was performed over a period of 36 h with sodium gluconate. After the lag phase samples were withdrawn periodically for the measurement of $OD_{550\,nm}$ and for ammonia assay. Figure 4-1 shows the cell growth, measured as $OD_{550\,nm}$ and nitrogen consumption as a function of time. Furthermore, the maximum reduction of pO_2 respectively the maximum increase of pCO_2 presents the maximal growth rate. The cell dry weight at the end of fermentation was 2.2 g/L and the limitation of nitrogen was reached after 20 h. There was no stationary phase observable due to the assumed intracellular PHB accumulation associated with a linear increase of the OD after exponential growth.

Fed-batch cultivation of *R. eutropha* was performed using crude glycerol as substrate. The reached PHB content of the bacteria depends on the pollution of the crude glycerol with salts, as a function of the caustic solutions and acids assigned to the soaping of rape oil and following neutralization (Mothes et al. 2007).

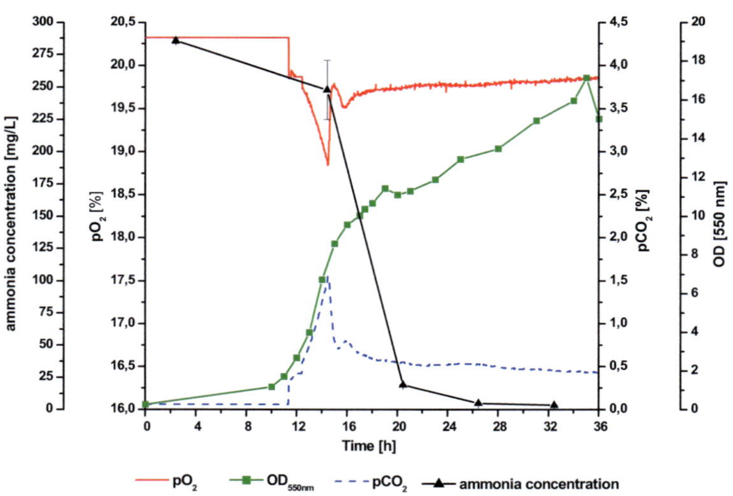

Figure 4-1: Growth of *R. eutropha* in a batch culture, measured as OD (■), pO_2 and pCO_2, and limitation of nitrogen, measured as ammonia concentration (▲)

Samples were periodically withdrawn for the measurement of cell dry weight, ammonia concentration, glycerine concentration and mass percentage of PHB. Figure 4-2 shows cell growth, measured as nitrogen consumption and cell dry weight as a function of time.

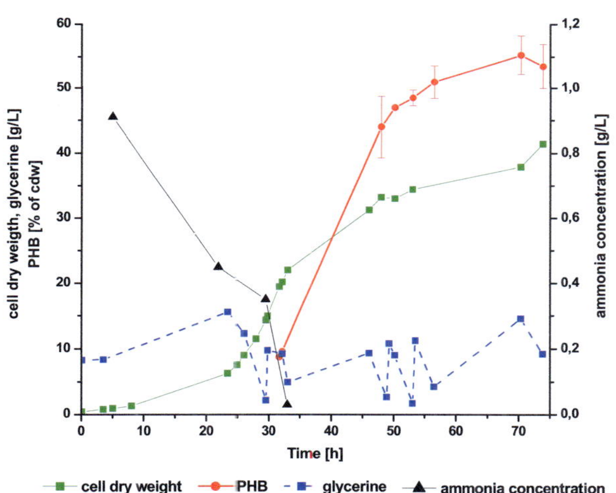

Figure 4-2: Growth of *R. eutropha* JMP134 (DSM 4058) in a fed-batch process on basis of glycerine measured as cell dry weight (■), mass percentage PHB (●), ammonia concentration (▲) and discontinuously fed glycerine

Furthermore, the discontinuously feed of the carbon source glycerine and the accumulation of PHB after nitrogen limitation are presented. The cell dry weight at the end of fermentation was 40 g/L and PHB concentration was about 50% of the residual biomass. Foam development in the fermenter, amount of substrate, pollution of the crude glycerol with fatty acids and ventilation speed are important factors affecting high PHB content. For larger fermentation scale, mechanical foam destruction is reasonable.

4.1.2 Electrofiltration of PHB dispersions

After cell disruption, electrofiltration was performed using the PHB rich lysates obtained from batch and fed-batch processes. The effects of applied pressure and electric field strength on the filtration kinetics were tested by varying both parameters. Additionally, during the experiments, process control investigations including variation of the same parameters were conducted in order to determine the optimum conditions. The t_f/V_f versus V_f diagrams were demonstrated for the interpretation of the filtration characteristics. The slope of the curves on figures Figure 4-3, Figure 4-4 and Figure 4-5 describes the specific filter cake resistances which is a product related characteristic.

Effect of different pressures and electric field strengths upon the efficiency of electrofiltration of PHB containing dispersions, was investigated by application of 1, 2 and 4 bars overpressure, with and without electric field. The filtrate mass values, measured online during the experiments, were used to create the t_f/V_f versus V_f diagrams.

The application of higher pressure in experiments with incompressible filter cakes results in an increased filtrate mass as described in Darcy's law (Darcy 1856). However, biopolymers are highly compressible molecules and the increase of pressure does not increase the filtrate mass proportionally. The effect of different constant pressures on filtration kinetics with no electric field is presented in Figure 4-3. As demonstrated from the similarity of the curves, applied pressure has no significant effect on filtration behavior for PHB rich lysates derived from batch and fed-batch cultivations. This result was based on the high compressibility of the product hindering high flow rates.

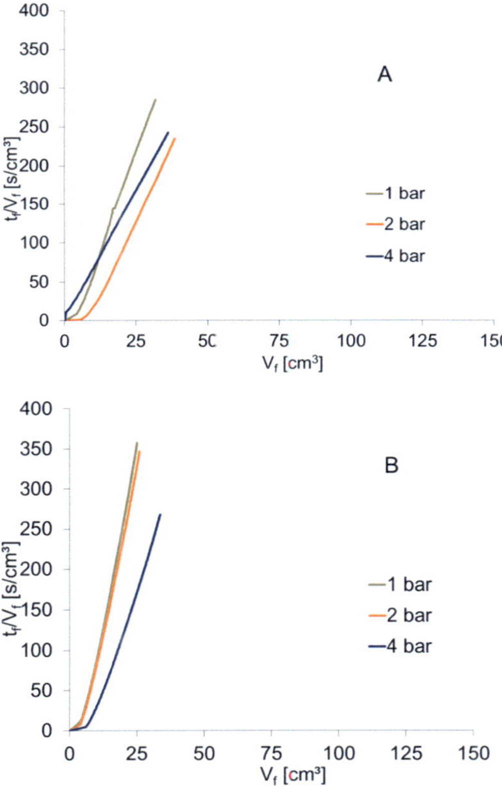

Figure 4-3: t_f/V_f versus V_f diagram of PHB rich lysates derived from batch (A) and fed-batch (B) cultivation at 0 V/mm with different pressures

Higher pressure resulted in compression of the PHB filter cakes increasing specific filter cake resistances. The linearity of the curves indicates the same course of specific filter cake resistance based on the experimental conditions with no electric field. The average filtrate fluxes of batch lysate were about 21 L/m^{2}h and the corresponding fluxes of fed-batch lysate were about 18 L/m²h independent of applied pressure. These values present the insignificant effect of higher pressures on filtration kinetics. Experiments without electric field at a pressure of 4 bars increased the product concentration in filter chamber in comparison to the initial concentration of PHB containing dispersion 3.5 times for both dispersions tested. Therefore, the limitations in filtrating PHB dispersions could not be overcome simply by increasing the pressure.

However, the application of an electric field leads to significant changes in curve progression of t_f/V_f versus V_f diagrams and relatedly specific filter cake resistances. The effect of different constant electric fields on filtration kinetics at a pressure of 4 bars pressure conditions is presented in Figure 4-4. Additionally, the experimental t_f/V_f versus V_f diagram compares the effects of different applied electric field strength on both PHB rich lysates obtained by batch and fed-batch cultivations. The selected conditions were based on the overall specifics of the products.

As a result of the migration of PHB granules towards anode side, a thin filter cake film on cathode side membrane was formed. A consequential significant reduction in specific filter cake resistances could be visualized by the slope of the curves. The graphical representation of the experiments with 0 V/mm demonstrated a steeper slope than those of 2 and 4 V/mm for PHB rich lysates. Due to the reduced specific filter cake resistances, filtrate fluxes of PHB rich lysates were increased in comparison to conventional filtration.

The filtrate fluxes of batch produced dispersions with application of 4 bars pressure were 53 and 87 L/m²h for 2 and 4 V/mm respectively. The corresponding fluxes of dispersions derived from fed-batch cultivation were 40 and 63 L/m²h for 2 and 4 V/mm respectively.

Concentration in the filter chamber for the elapsed time, after experiment with conditions of 4 bars and 4 V/mm, was increased 15 and 12 times for batch and fed-batch processes respectively in comparison to initial concentration of PHB rich lysates. While a final filtrate mass of 36 g was achieved after 150 min of batch lysate filtration without an electric field, the same mass was obtained after only 14 min by applying an electric field of 4 V/mm representing a time saving of approximately 90%.

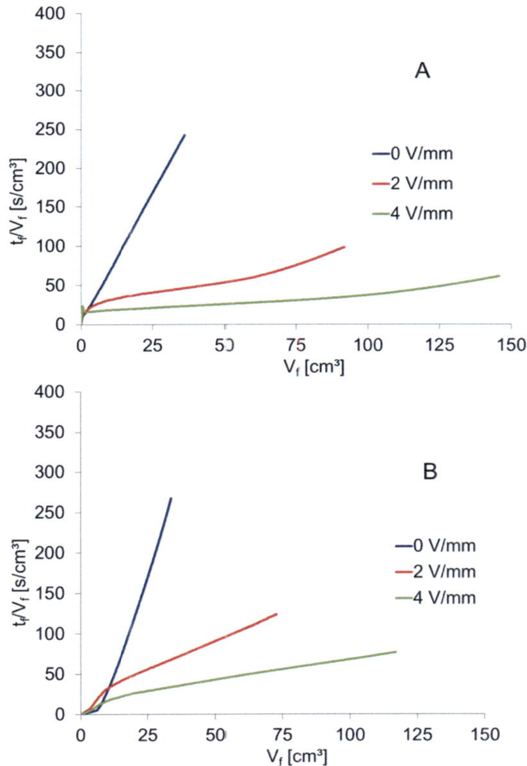

Figure 4-4: t_f/V_f versus V_f diagram of PHB rich lysates derived from batch (A) and fed-batch (B) cultivation at 4 bars with different electric field strength applied

For fed-batch filtration the same phenomenon was observed. Therefore, the applied electric field had a definitive positive effect on the filtration kinetics of PHB dispersions. The results demonstrated that the application of electric field in the filter chamber improved the filtration kinetics significantly. Both PHB dispersions have similar filtration profile which could be justified by the comparison of the diagrams of the applied processes. The resemblance was based on the similar zeta potentials values. Despite the comparable filtration profiles, there is a slight difference between the specific filter cake resistances represented by the slope of the curves in Figure 4-3 and Figure 4-4. The ones resulting from PHB dispersions derived after fed-batch process were steeper than those of produced by batch process. The results emerged from the separate cultivation procedures used and the different molecular weights of the products. Higher molecular weights can cause an increase in both filter cake resistances and hydrodynamic radius of the particles.

In order to investigate the influence of parameter changes on the filtration kinetics during running process, process control experiments with fed-batch produced dispersion were performed. Variation of pressure and electric field strength during the experiments were used to optimize the process. In Figure 4-5, the t_f/V_f versus V_f plot, resulting from this optimization, was compared with an experiment with constant applied pressure and electric field strength (4 bars and 4 V/mm). During the process control experiment, the electric field strength was increased in intervals starting from 1 V/mm at 1000, 2000 and 4000 s of the experiment with 1 V/mm increments. The time interval division in the Figure 4-5 is visualized by the sections I, II, III and IV.

Pressure was increased in each interval as follows: for the first section 1 bar, for the second and third sections 2 bars and for the fourth section 4 bars. After 4000 s the conditions were identical to the experiment with the constant applied pressure and electric field strength.

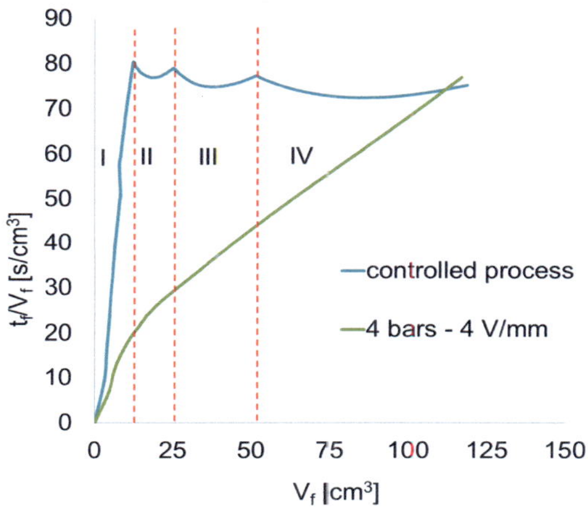

Figure 4-5: t_f/V_f versus V_f diagram of fed-batch derived PHB dispersions with variable and constant parameters

The first step of the process with variable parameters is characterized by the steep profile of the curve corresponding to a high specific filter cake resistance. The applied electric field (1 V/mm) was not high enough to influence the mobility of the charged molecules – conditions similar to the ones without electric field. The specific filter cake resistance in the first section demonstrates a higher value in comparison to the following steps and as a result, the filtrate flux during the first step was lower. The increase in electric field strength in section II induced the significant reduction in specific filter cake resistance. The described particle migration is a result of the application of electric field, which changes the behavior of the molecules. This effect was further verified by the nearly constant progress of the specific filter cake resistance in sections III and IV. Although filtration kinetics is mainly determined by the application of electric field, the applied pressure also contributes to the filtration process. Around 8000 s where the filtrate mass reaches 110 mL, both curves intercept each other. The filter cake formed after the process with varied process conditions has lower specific filter cake resistance compared to the one resulting from the process with constant conditions. This particular characteristic facilitate the higher amount of filtrate mass for the process control experiment after the crossing point between two processes in section IV of Figure 4-5.

The common purification of PHB requires the use of organic solvent such as chloroform and precipitation with hexane. Addition of large amounts of organic solvents is hazardous to the environment and increases the production costs. In the case of hexane is evaporated after precipitation, 36500 J/100g additional energy is required (Green and Perry 2008). The multiple use of solvent during purification increases the energy consumption.

In comparison, the energy required for the electrofiltration of the PHB dispersion was 7200 J and 10800 J per 100 g filtrate, derived from batch and fed-batch process with constant conditions, respectively. Electrofiltration required 70-80% less energy compared with the common method. The difference in energy levels is based on the product specifics, arising from the different synthesis conditions, and eventually influencing the filtration behavior. The energy required for the process control electrofiltration experiments which was performed with fed batch produced dispersion was 8900 J/100 g filtrate (%17 less than the experiment with constant conditions). The results demonstrate that the improved purification technology has high optimization potential leading to reduction of energy consumption in the production technology of PHB by electrofiltration.

4.1.3 Characterization of separation of PHB

Zeta potential

For a better evaluation of the efficiency of the experiments, prior electrofiltration, determination of zeta potential of batch and fed-batch derived granules, was performed and presented as a function of pH in Figure 4-6. High magnitude of zeta potential corresponds to high stabi ity of the dispersion and high molecule charge. The charge of the particles, depending on ionic strength and pH, influences the distribution of surrounding ions.

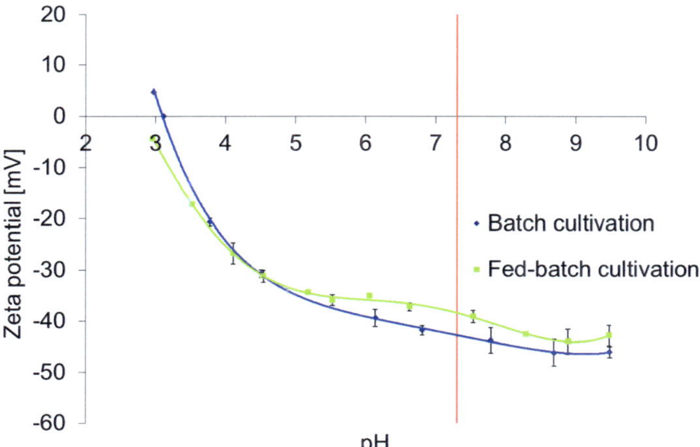

Figure 4-6: Zeta potential of PHB samples derived after batch and fed-batch processes

Based on the similar ionic strength values of the lysates, profile of the curves in Figure 4-6, suggests that zeta potential does not depend on the cultivation method (batch and fed-batch). The net electrical charge of the PHB granules reaches zero around pH 3 designated as isoelectric point and increases with the increase of pH. The experiments were conducted at pH 7.3 considering the original pH of the samples where both lysates have a zeta potential value of approximately -40 mV.

Molecular weight

GPC analysis of each PHB lysate and the purchased PHB sample as a reference was performed in order to determine the molecular weights of PHB samples. The molecular weight of PHB usually ranges between 10 and 3000 kDa (Fiorese et al. 2009) and GPC analysis revealed notable differences in the samples tested. A molecular weight of 129 kDa was obtained through batch cultivation with gluconate as carbon source while fed-batch cultivation based on glycerine resulted in polymer with average molecular weight of 357 kDa. The results confirm that the molecular weight of PHB can differ with carbon source and cultivation conditions (Rebah et al. 2004). In addition, for electrofiltration experiments, molecular weight is an important characteristic, affecting the specific filter cake resistances and hence filtration kinetics.

PHB content

Due to the insufficient filter cake thickness, the GC/MS analysis was performed using dispersions from the inside of the filter chamber. The analysis was necessary in order to investigate influence of the purification technique upon PHB content. Analysis prior electrofiltration experiments revealed PHB content in the dispersion after cell disruption. The values ranged between 28-30% for the dispersions obtained after batch and fed-batch processes.

As discussed in section 4.1.2, the pressure has low influence on the filtration kinetics compared to the electric field. Therefore, GC/MS investigations were focused on the effect of electric field with constant intermediate pressure of 2 bars. For experiments with 0 V/mm the determined PHB content remained within the initial range of the starting dispersion. The lack of electric field hindered the intensive formation of filter cakes and as a result PHB content had no significant changes. However, as demonstrated from the previous results, the application of electric field had notable effect on the filtration kinetics and consequently changed the composition of the solution remaining in the filter chamber. After experiments with 4 V/mm PHB content in the filter chamber was reduced to 13.5% for both processed dispersions. The reduction in specific filter cake resistances by application of electric field increased the filtrate flux, intensified the cake formation on the anode side membrane and eventually reduced the PHB content in the dispersion.

Structural integrity

Raman spectroscopy was used to detect the possible structural changes caused by the applied electric field. Obtained spectra revealed the presence of different organic functional groups that represent the main macromolecular building blocks (nucleic acids, proteins, carbohydrates and lipids) as previously described by Naumann (2001). The Raman spectra of the lysates before and after electrofiltration at 4 bars with 0 and 4 V/mm are presented in Figure 4-7. All other spectra obtained by using different process conditions were similar. The figure also contains the spectrum of the purchased PHB.

Despite the different origins of the analysed samples regarding particle size and sample matrix, characteristic bands, identical to the standard containing 99% PHB, were detected in all samples. The additional peak occurring between 1680 and 1650 cm^{-1} in batch and fed-batch derived samples represents Amid I band. This peak corresponds to the double bond between carbon and oxygen in the carbonyl group of peptide bonds and therefore results from amides, respectively proteins that are parts of the PHB granules. Only insignificant differences among the samples were detected presumably related to impurities and differences in particle sizes. Results confirm that no structural changes or damages of PHB in the analysed samples were identified by Raman spectroscopy.

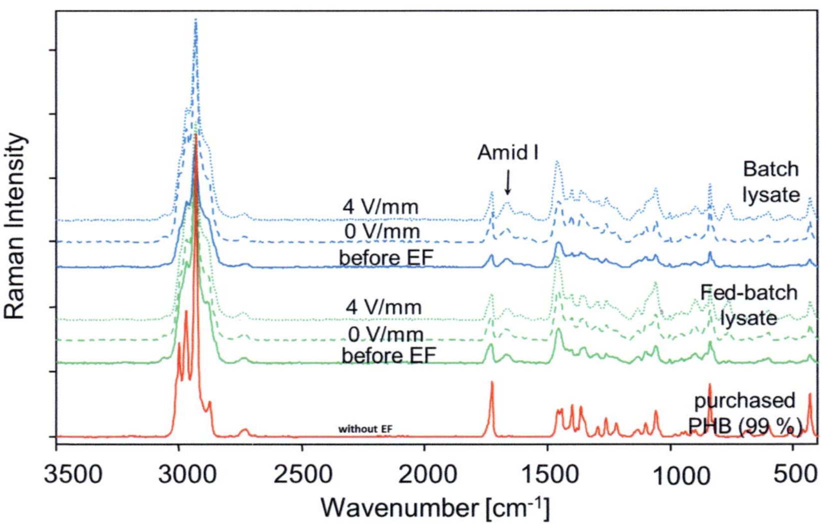

Figure 4-7: Raman spectra of PHB samples under different filtration conditions

This research presents an innovative approach for the purification of PHB and introduces electrofiltration as perspective method in the production chain of biopolymers. The application of different cultivation strategies including different media, strains and conditions affected the structural and dynamical properties of PHB dispersions causing the change in filtration behavior. In comparison to other purification methods electrofiltration demonstrated advantages concerning its high performance and cost efficiency.

The lack of unwanted modifications of the final product, as revealed by the characterization experiments facilitates the direct practical applications of PHB, opening new possibilities for the bioplastic industry. The results confirm that the optimization and scale-up potential of the microbiological PHB production meets the current downstream processing trends to reduce the number of steps and apply integrated technologies.

4.2 Separation of chitosan

4.2.1 Microbial production of chitosan

The microbial chitosan production was initiated by the cultivation of both fungi studied under batch and repeated batch conditions. An illustration of a typical batch process involving *M. rouxii* is presented in Figure 4-8. The strong increase of extractable chitosan during the growth phase was attributed to free chitosan molecules, due to the active growth of hyphal tips (Tan et al. 1996). In the cell walls of Zygomycetes, chitin and chitosan are in a dynamic balance. The transition into the stationary growth phase leads to the conversion of more chitosan to chitin and other cell wall components. Therefore, the precise moment of harvesting is essential.

In the experiments with *M. rouxii* and *A. coerulea*, the maximum chitosan productivity was obtained between 14-16 hours of cultivation. This corresponds to previous work with other fungi reported by Shimahara et al. (1989) and Tan et al. (1996), where the maximum chitosan yield of

Absidia butleri and *Rhizopus oryzae* was confirmed during the late exponential growth phase.

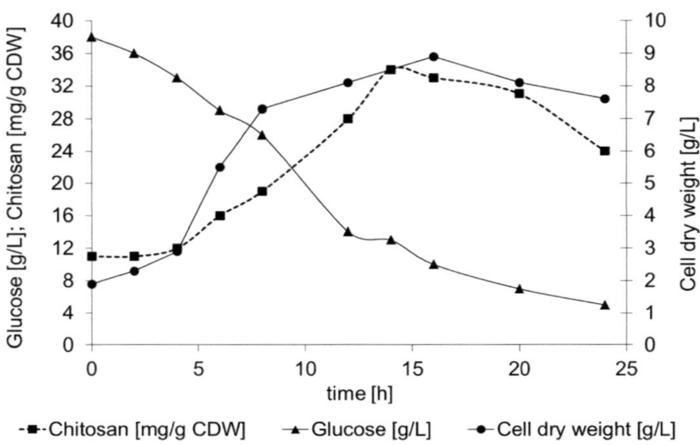

Figure 4-8: Bioreactor cultivation of *M. rouxii* under batch conditions and the formation of chitosan

In addition to the physiological state, the chitosan yield from fungal mycelia is also affected by the extraction method. Extraction experiments have revealed that the use of hydrochloric acid instead of acetic acid has advantages in terms of the final yield (White et al. 1979), but due to the higher degree of deacetylation of the extracts and in order to avoid changes in the degree of polymerization, the acetic acid method was preferred (Synowiecki and Al-Khateeb 1997).

The alkali-insoluble material (AIM) was about 25% of the total dry biomass of Zygomycetes. The amount of extracted chitosan obtained from 100 g of dried biomass was 4.2 and 5.6 g for *M. rouxii* and *A. coerulea*, respectively. This result is comparable with previously reported chitosan yields from *M. rouxii* (Trutnau et al. 2009).

4.2.2 Electrofiltration of chitosan dispersions

The effects of applied pressure and electric field on the filtration kinetics of chitosan dispersions were tested individually by varying both parameters. The objective of these experiments was to demonstrate the advantages of electrofiltration of chitosan over conventional dead-end filtration. To investigate the influence of different pressures upon the efficiency of electrofiltration, separate experiments with 1, 2 and 4 bars overpressure, with and without and electric field, were conducted with chitosan dispersions in a filter chamber. The mass of filtrates were measured and these values were used for the calculation of the corresponding filtrate fluxes.

The comparison between the filtrate mass versus filtration time plots, obtained by electrofiltration of chitosan dispersions subjected to a constant electric field of 4 V/mm and the three different pressures is presented in Figure 4-9. The increase of applied pressure improves filtrate flux. However, this effect is not directly proportional to pressure due to the compressibility of the filter cake which increased specific filter cake resistances. In the experiments, the filtration kinetics of chitosan dispersions revealed that higher pressure increased the filtrate flux, mostly in processes with dispersion of crab chitosan, with lower effects on the samples from *M. rouxii* and *A. coerulea* (Table 4-1). The concentration factors based on the permeate mass of samples at 4 bars were 12, 6 and 4 for crabs, *M. rouxii* and *A. coerulea* derived dispersions, respectively. The different behavior in Figure 4-9 is probably based on the different electrostatic repulsion and interactions between polymers which can influence the compressibility of chitosan filter cakes.

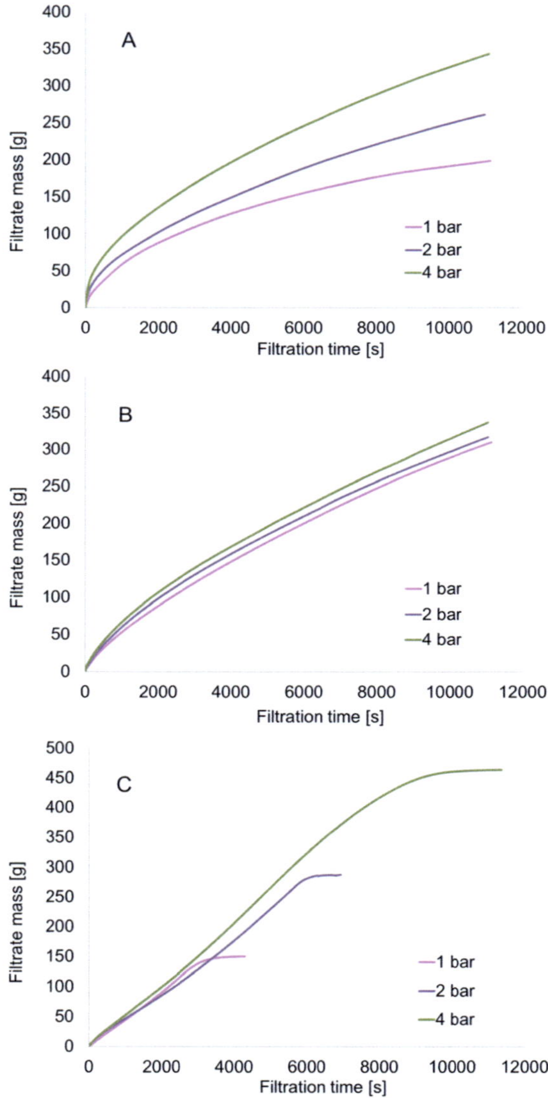

Figure 4-9: Time dependent comparison of filtrate mass of chitosan dispersions for electrofiltration experiments at constant electric field of 4 V/mm. Chitosan samples were derived from crabs (A), *M. rouxii* (B) and *A. coerulea* (C)

The production procedure of fungal chitosan led to higher conductivity of the product (1500 µS/cm) compared to crab chitosan (620 µS/cm), which was commercially avai able in a pure state.

Table 4-1: Average filtrate f ux of electrofiltration experiments undei 0 V/mm with 1, 2 ard 4 bars conditions

	Filtrate flux $[L/m^2h]$		
	Crab	**M. rouxii**	**A. coerulea**
1 bar	41	17	15
2 bar	52	21	16
4 bar	67	27	17

Higher conductivity resulted in screening of the charge of chitosan molecules by salt ions which originate from the cultivation medium. The phenomenon can be explained by DLVO theory which defines the balance between attractive (van der Waals) interactions and electrostatic repulsion (Hidalgo-Alvarez et al. 1996). As a result of increased ionic strength, electrostatic repulsion between chitosan molecules decreased and compression of filter cake altered the structure of the filter cake, leading to a decrease in porosity, which consequently increased the specific filter cake resistance. Specific technological difficulties characteristic for the filtration of chitosan dispersions are due to the high compressibility of filter cakes. The results presented here demonstrate that those difficulties could not be overcome simply by increasing the pressure.

The filter cake of chitosan from *A. coerulea* was characterized by the highest compressibility which could be recognized by the lack of pressure influence. Further confirmation of this phenomenon was the observed end of filtration, visualized by the stable filtrate amount in

Figure 4-9C, characteristic only for the experiments with *A. coerulea* with application of an electric field. The increasing of pressure resulted in a more compressed structure of the filter cake. Therefore, a higher chitosan concentration in the filter chamber was required until the cathode and anode filter cakes contacted with each other, causing the end of filtration. In addition to this effect, the application of an electric field increased the migration velocity of chitosan molecules according to Manning theory (Manning 1981), and the formation of the cake was faster compared to tests without an electric field. Additional application of a stronger electric field resulted in earlier filtration stop at lower pressures. Processes with *M. rouxii* and crab-derived chitosan dispersions could not reach the end of filtration in the experimental time tested. This effect suggests a high optimization potential towards more effective and economical purification of chitosan dispersions.

To investigate the influence of different electric field strength upon the efficiency of electrofiltration continuous electric fields of 2 V/mm and 4 V/mm, together with a reference experiment with 0 V/mm, were applied under different constant pressure conditions. Figure 4-10 compares the effects of different electric fields upon electrofiltration of all three chitosan dispersions as a t_f/V_f versus V_f diagram under 4 bars condition, selected according to the overall specifics of the products. The t_f/V_f versus V_f diagrams are valuable representatives of the filtration characteristics (Weber and Stahl 2002). The slope of the curves in the t_f/V_f versus V_f plot on Figure 4-10 describes the specific chitosan filter cake resistance which is a product related characteristic. The curve corresponding to 0 V/mm demonstrates a steeper slope than those of 2 and 4 V/mm for all chitosan dispersions.

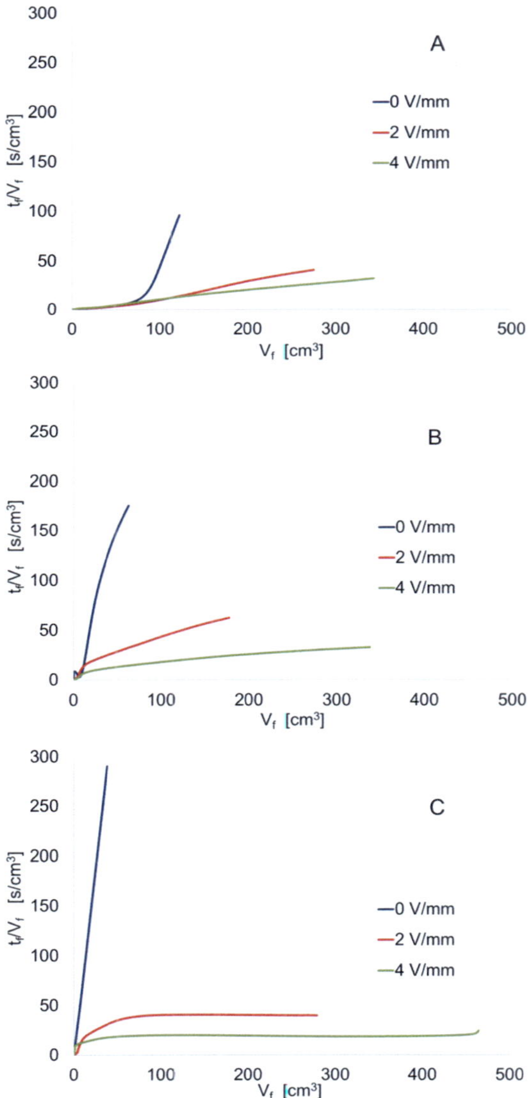

Figure 4-10: t_f/V_f versus V_f diagram of chitosan dispersions derived from crabs (A), *M. rouxii* (B) and *A. coerulea* (C) at 4 bars with electric field strengths of 0, 2 and 4 V/mm.

This correlation indicates that with an increase to the electric field strength, the chitosan filter cake resistance decreased significantly, leading to a higher permeate mass and filtrate flux. Therefore, the applied voltage had a strong positive effect on the filtration kinetics of chitosan dispersions.

Increasing steepness of the curves for experiments with 0 V/mm, confirmed the change in specific filter cake resistance which increased from crab-derived through *A. coerulea*-derived chitosan. With no electric field applied, higher specific filter cake resistances of fungal chitosan are a result of the salt content of the dispersions and lead to lower filtrate flux (Table 4-2). Application of an electric field overcomes the limitations induced by high specific filter cake resistances.

Table 4-2: Average filtrate flux of electrofiltration experiments under 4 bars with 0, 2 and 4 V/mm conditions

	Filtrate flux [L/m^{2}h]		
	Crab	**M. rouxii**	**A. coerulea**
0 V/mm	67	27	17
2 V/mm	127	75	90
4 V/mm	151	135	181

The concentration factors based on the permeate mass of samples at 4 V/mm were 34, 33 and 46 for crabs, *M. rouxii* and *A. coerulea* derived dispersions, respectively. The difference in filtrate flux values of chitosan dispersions were influenced by the different zeta potentials. Chitosan derived from crabs, *M. rouxii* and *A. coerulea* have a zeta potential of +61, +42 and +46 mV at a pH of 3, respectively. The lower zeta potential of fungal chitosan dispersions is based on their higher ionic strength compared to crabs derived chitosan. High salt concentration of fungal chitosan caused decreasing of the electrostatic repulsions between molecules, leading to a decrease in zeta potential.

Another important factor determining the effectiveness of our electrofiltration experiments, despite the relatively lower zeta potential of fungal chitosan, was the concentration of acetic acid. The higher acetic acid content in the fungal chitosan dispersions (up to 10%) intensified the filtration kinetics with application of electric field. The difference in the filtrate flux of both fungi derived chitosan dispersions could be related to protein content. The lower amount of protein content in *A. coerulea*-chitosan corresponds to higher filtrate flux, presumably due to the relatively unrestricted movement of chitosan molecules.

As these experiments demonstrate, electrofiltration leads to increasing filtrate fluxes and reduces the processing time. For the same amount of filtrate, electrofiltration was 7 times faster than conventional filtration for chitosan from crabs, 12 times faster for *M. rouxii*-derived chitosan and 15 times faster for *A. coerulea*-derived chitosan.

The advantages of electrofiltration are demonstrated not only by the values regarding its efficiency but also by the formation of the final product presented in Figure 4-11. The images represent well-structured cathode side chitosan filter cakes. The difference in color between the three final products arises due to an effect from different components in the cultivation medium, leading to a darker appearance in the fungus-derived chitosan. Additionally, the color change could be attributed to the occurrence of a Maillard reaction during sterilization of the medium.

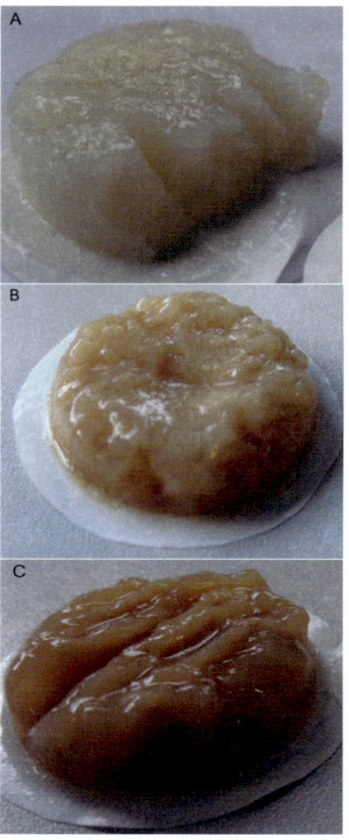

Figure 4-11: Formation of filter cakes after electrofiltration of chitosan dispersions derived from crabs (A), *M. rouxii* (B) and *A. coerulea* (C) under 4 bars and 4 V/mm conditions

Electrofiltration is a hybrid technology which fulfills the main objective of the current research trends in the field of downstream processing, namely to shorten and simplify the entire process by combining different approaches. This results in a reduction in production costs. The economic efficiency of electrofiltration was estimated from the energy balance at the end of the process and compared with purification steps of analogous techniques. Conditions of 4 bars and 4 V/mm were assumed for the balance in order to recreate the maximal possible consumption. The energy requirement for the electrofiltration of crab-derived chitosan was 10500 J/100 g filtrate, while for *M. rouxii* and *A. coerulea*-derived chitosan this was around 43000 J/100 g filtrate (Gözke et al. 2011). The standard precipitation approach requires centrifugation at the end of the process, consuming 120000 J for the same dispersion amount. Electrofiltration required 91% less energy for the separation of crab-derived chitosan and 64% less energy for the separation of fungus-derived chitosan. The different energy requirements are mainly based on the different conductivities and respective currents. The presented model values are only a demonstration of the potential advantages of electrofiltration and could be further reduced by applying lower pressures, for example in the cases of fungal chitosan purifications, or by applying technically optimized devices.

4.2.3 Characterization of separation of chitosan

In order to analyze the applicability of the developed chitosan purification strategy and to evaluate its effectiveness, detailed process and product characterizations were conducted. Possible changes in the chemical structure in the final product, arising from the application of an electric field and pressure were investigated and comparisons were made with the standard technique (precipitation) regarding the quality of chitosan

Degree of deacetylation (D_{DA})

The D_{DA}, one of the most important quality parameters characterizing chitosan, was determined after purification by means of electrofiltration. It was demonstrated that different electric fields and pressures had no negative effect on D_{DA}. This quality was in the range of 76-80% for chitosan from all sources, which meets the requirements for medical applications.

Viscosity and average molecular weight

The results demonstrate that fungal chitosan exhibited lower viscosity and molecular weight. The determined viscosity of dispersions derived from crabs, *M. rouxii* and *A. coerulea* were 4.37, 1.71 and 1.31 mPas, respectively. A tendency similar to the viscosity was observed for the molecular weights. By means of SEC-MALLS, the molecular weight of crab chitosan was determined to be 167 kDa, and was 96 kDa for *M. rouxii*-derived chitosan and 57 kDa for *A. coerulea*-derived chitosan. Low molecular weight is advantageous for further chitosan applications in medicine and also in agriculture, due to its ability to induce plant growth more effectively, in comparison to chitosan originating from crustaceans (Nge et al. 2006; Ondruschka et al. 2008).

Protein content

Protein content is an important parameter for applications of chitosan, especially in the field of medicine. For example, higher protein concentrations may cause inflammatory reactions. A high quality chitosan is characterized by protein content below 1% (Chatterjee et al. 2005). A comparison between protein contents (amount of g protein per g dry weight in terms of percent) in the filter cakes after electrofiltration (4 bars-2 V/mm) and precipitation is presented in Figure 4-12. The results clearly demonstrate the advantage of electrofiltration as a purification technique. For all three products tested, electrofiltration led to a notable reduction in the protein content compared with the standard precipitation technique.

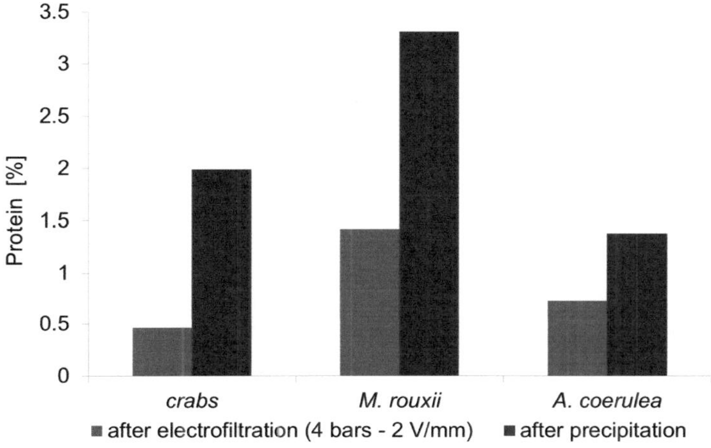

Figure 4-12: Comparison of protein content in chitosan filter cakes after electrofiltration (4 bars-2 V/mm) and after precipitation

The differences in protein content between the three chitosan dispersions tested were origin and preparation method related (Aranaz et al. 2009). The characteristics of chitosan from all sources fulfill the application requirements. Only the origin *M. rouxii* is slightly above acceptable levels. The results support electrofiltration as an advantageous alternative with industrial relevance.

Dry weight

Dry weight concentrations were determined from the final filter cakes of chitosan samples and the concentration factors were calculated from initial concentrations of all dispersions. A clear demonstration of the efficiency of the approach is visualized in Figure 4-13. For all the chitosan dispersions tested, the application of an electric field increased the dry weight concentration factors for each constant pressure.

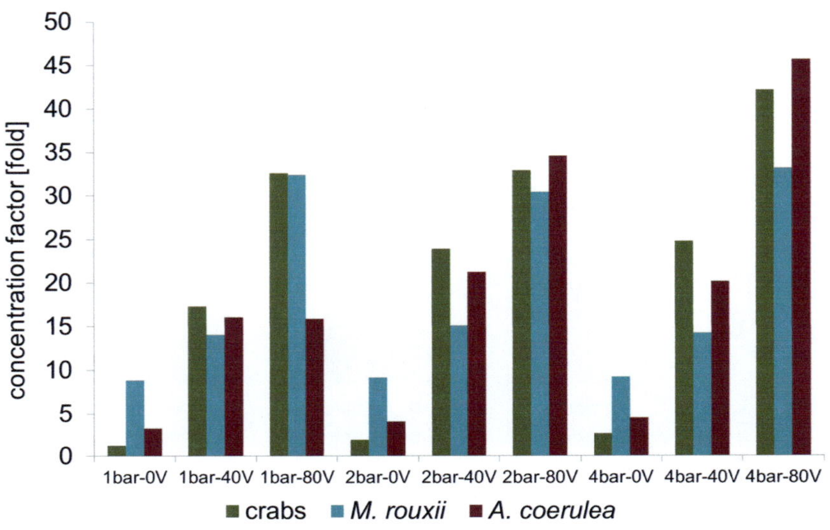

Figure 4-13: Dry weight concentration factors of chitosan filter cakes obtained after electrofiltration experiments under various pressures and electric field strengths for the crabs, *M. rouxii* and *A. coerulea* derived dispersions.

The only exception from this result was the concentration factor of the filter cakes from *A. coerulea* between 40 and 80 V at 1 bar, which in Figure 4-13 presents almost equal values. This phenomenon is based on the correlation between the strength of the electric field and the duration of the process at lower pressures as described previously.In comparison to conventional filtration, dry weight concentration factors increased from 3, 9, 5 to 42, 33, 46 for crabs, *M. rouxii* and *A. coerulea*, respectively. The increase of pressure in the absence of an electric field led to a negligible change in dry weight concentration factors due to the compressibility of the filter cakes, a previously discussed effect. The increase of pressure at constant electric fields had a similar effect for chitosan from crabs and *M. rouxii*. For the experiments with *A. coerulea*, the changing dry weight concentration profile was affected by the different compressibility characteristics. The different behavior of all chitosan dispersions was derived from the different effects of the electric field upon the processes which were predetermined by the origin of the product.

Heavy metals

The content of heavy metals, an important characteristic for possible applications, was determined in all samples. The amounts of Pb, Cd and Cu were below the detection limits of the method. Therefore, only the content of Zn in all samples before and after electrofiltration was determined.

The results demonstrate that electrofiltration significantly reduced the amount of Zn in the final product. The initial Zn concentrations for crabs (0.23 mg/L), *M. rouxii* (0.16 mg/L) and *A. coerulea* (0.32 mg/L) were reduced after electrofiltration to 0.012, 0.045 and 0.060 mg/L, respectively. Differences in the efficiency of Zn purification for the final products arose from the initial metal concentrations. For fungi-derived chitosan, the metal originated from the medium components and for crustaceans from the heavy metal content of sea water.

FT-Raman spectroscopy

Another important characteristic regarding the application of the product is the chemical structure of chitosan. In order to detect possible pressure and electric field-related damage, all final products were analyzed using FT-Raman spectroscopy. Due to the common appearance of fluorescence in biological samples which strongly interferes a conventional Raman measurement, FT-Raman spectroscopy was used.

The application of a near-IR laser (λ = 1064 nm) in this technique reduces the appearance of fluorescence nearly down to zero and provides spectra of a good quality.

Several results illustrating the characteristic spectra of chitosan samples derived from experiments with 4 bar-0 V/mm, 4 bar-2 V/mm, 4 bar-4 V/mm are presented in Figure 4-14 and the following bands could be detected (Table 4-3):

Table 4-3: Chitosan characteristic bands detected by FT-Raman analysis

	Wavenumber [cm^{-1}]	Assignment
1	2900-2940	C-H
2	1630-1660	N-H
3	1370-1390	-CH$_3$
4	1050-1150	-OH

The FT-Raman spectra of the three chitosan samples were comprised of broad bands involving mainly CH, NH_2, CH_3 and OH vibrations (Vasconcelos et al. 2008; Zhang et al. 2010). A typical and intense band (label 1) resulting from C-H stretching occurred at around 2900-2940 cm^{-1}. This peak was due to the organic substrate of the chitosan. Another characteristic band for the NH_2-group occurred at 1630-1660 cm^{-1} (label 2). Label 3, representing the CH_3-group at 1370-1390 cm$^{-1,}$ and label 4 show the OH-vibrations at 1050-1150 cm^{-1}. The intensity of the bands was related to the amount of functional groups and could vary, depending on the color and the crystalline structure of the lyophilized samples.

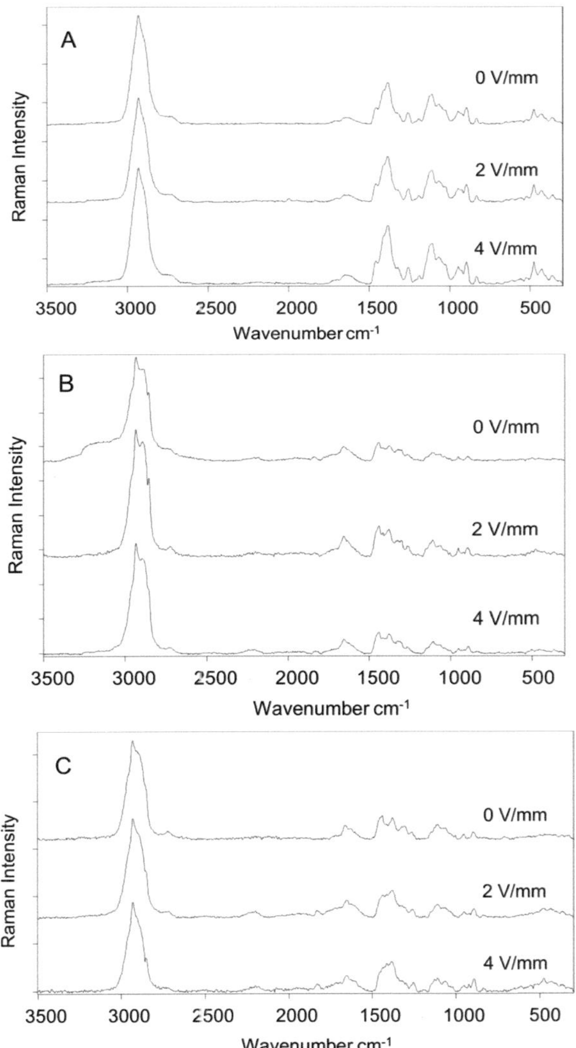

Figure 4-14: FT-Raman spectra of filter cakes after electrofiltration of chitosan dispersions from crabs (A), *M. rouxii* (B) and *A. coerulea* (C) under 4 bars with 0, 2 and 4 V/mm conditions.

The FT-Raman spectra confirmed the absence of undesired structural changes as a result of the applied electric field during the purification of chitosan samples. Additionally, according to data not presented, pressure had no significant influence either.

This research describes an alternative approach in the downstream processing of chitosan and lists the advantages of the technique that characterize it as an appropriate solution for industrial scale chitosan production. The introduction of electrofiltration is particularly attractive for the purification of chitosan, which is difficult to filtrate due to its high compressibility. It can reduce processing time and lower the number of purification steps, decreasing the total production costs. The experiments delivered excellent results concerning the process parameters: increasing filtrate flux, optimizing the applied conditions (pressure and electric field) and improving the product characteristics. The comparison between the three products tested revealed a strong dependency between the product's origin and the technological conditions required. The application of high quality fungal chitosan, which, regarding its characteristics is particularly interesting for the medical industry, requires further optimization. Our experiments overcome many of the technological challenges of the previously used methods and the results obtained open new possibilities for the downstream processing of biopolymers.

4.3 Separation of hyaluronic acid

4.3.1 Electrofiltration of hyaluronic acid dispersions

The effect of various process parameters upon the efficiency of electrofiltration of hyaluronic acid dispersions was investigated by conducting separate experiments with 1, 2 and 4 bar overpressure, with and without applied electric field. In addition, to determine the influence of specific buffer solutions on filtration kinetics several experiments with various buffer compositions were compared. Based on this result, further experiments were performed.

The t_f/V_f versus V_f diagrams were created in order to interpret the filtration characteristics based on the slope alteration of the curves, describing the specific filter cake resistances (Figure 4-15). The standard progress of electrofiltration experiments has 3 distinct sections (see section 2.4.2) divided according to the changes in specific filter cake resistances. In the first section due to the higher effect of hydrodynamic resistance force in comparison to the electrophoretic force, the specific filter cake resistance is high. After the equilibrium of the forces, no further filter cake is formed on the cathode side. In the third section, as a result of the constant accumulation of product on the anode side membrane, the filter cake formed reaches the cathode side filter film, increasing the specific filter cake resistance.

All curves on Figure 4-15, representing specific filter cake resistances exhibit almost the same tendency in the first two sections of the curves. However, the experiments using buffer of acetic acid-sodium acetate and potassium hydrogen phthalate-NaOH resulted in a higher specific filter cake resistance than the experiment with the buffer solution potassium dihydrogen phosphate and dipotassium hydrogen phosphate.

In the same time, the use of phosphate buffer led to higher permeate mass. Therefore for the further experiments, the latter buffer system was used. The achieved higher permeate mass is considered to be related to the molecule/ion interactions through the membranes. The inorganic composition of the buffer solution prevents the corrosion of the electrodes. Series of experiments demonstrated that different buffer solutions have different influences on filtration kinetics.

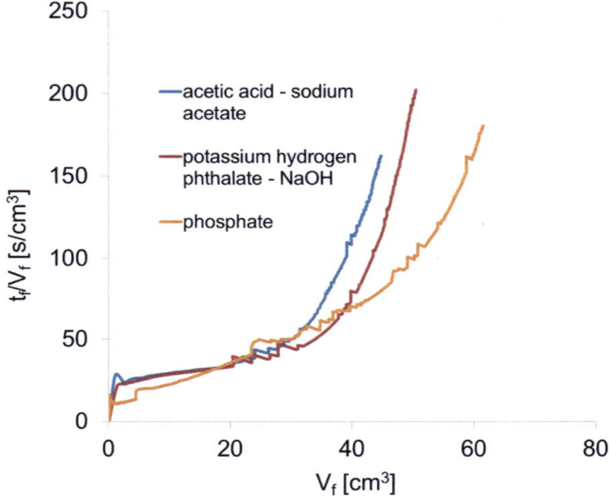

Figure 4-15: t_f/V_f versus V_f diagram of hyaluronic acid dispersions under conditions of 2 V/mm and 1 bar with various buffer solutions used

Filtration capacity of hyaluronic acid dispersions is limited, based on the compressibility characteristic of the biopolymer. The effect of pressure on filtration kinetics of hyaluronic acid dispersions is demonstrated on Figure 4-16. In general, the application of higher pressures results in increased specific filter cake resistances. In the case of no electric field applied, the hydrodynamic resistance force is the only force that determines the motion of the molecules and the height of filter cakes on both sides is equal. The applied pressure results in compression of the filter cake depending on the compressibility of the product. Increasing of the pressure leads to a delay in the contact of both filter cakes due to the compressed product. Figure 4-16 reveals that the experiment with 1 bar results in lower filtrate mass compared to the ones with 2 and 4 bars applied. Within the experimental frame, the differences between the permeate masses obtained with 2 bars and 4 bars are insignificant. However, a continuation beyond filtration time (in this case 12000 s) could lead to further change of the profile.

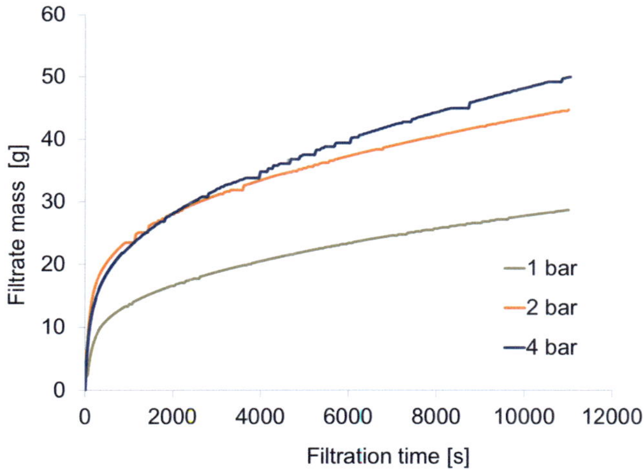

Figure 4-16: Filtrate mass versus filtration time plots of hyaluronic acid dispersions at 0 V/mm with different applied overpressures

Subsequent to the experiments concentration factors based on the permeate mass of samples were calculated. For 1, 2 and 4 bars conditions, these values were 2.8, 4.5 and 5 respectively and the obtained corresponding average filtrate fluxes were determined to 15, 23 and 25 L/m^{2}h.

Hyaluronic acid contains carboxylic groups determining its negative charge and zeta potential of -37 mV at pH 6. Experiments revealed promising results considering the filtration technology of hyaluronic acid dispersions. The charge facilitates the processing of the biopolymer by means of electric field. The effect of introduction of electric field is demonstrated in Figure 4-17 and compared with conventional filtration experiment.

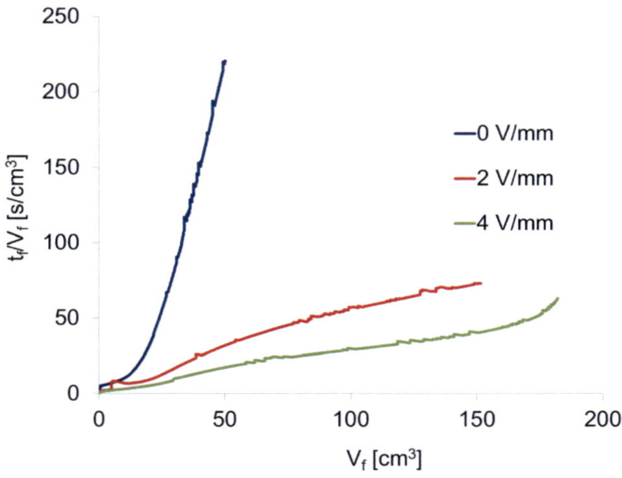

Figure 4-17: t_f/V_f versus V_f diagram of hyaluronic acid dispersions at 4 bars with different applied electric field strengths

The application of electric field leads to a decrease in specific filter cake resistances as seen from the slopes of the curves. The experiment with no electric field is characterized by a steeper curve compared to the ones derived after application of electric field of 2 and 4 V/mm.

The results demonstrate the high positive effect of electric field on the filtration kinetics of hyaluronic acid. This was based on the electrophoretic migration of negatively charged hyaluronic acid molecules directing them towards anode side membrane. Cathode side membrane remained almost free of filter cake, allowing the flow-through of the filtrate easily. For 0, 2 and 4 V/mm conditions the concentration factors at 4 bars were 5, 15 and 18 respectively and the obtained corresponding average filtrate fluxes were determined to 22, 60 and 80 L/m^2h for the elapsed time. The experiment follows the already described theoretical division of the electrofiltration process into sections. However, only the experiment with 4 V/mm reached the third section where the specific filter cake resistance increases. The composition of hyaluronic acid after electrofiltration (conditions 4 bars and 4 V/mm) is presented in Figure 4-18. The product is clear with a high viscosity. The energy need was 8100 J/100 g filtrate as a comparable value with other investigated products.

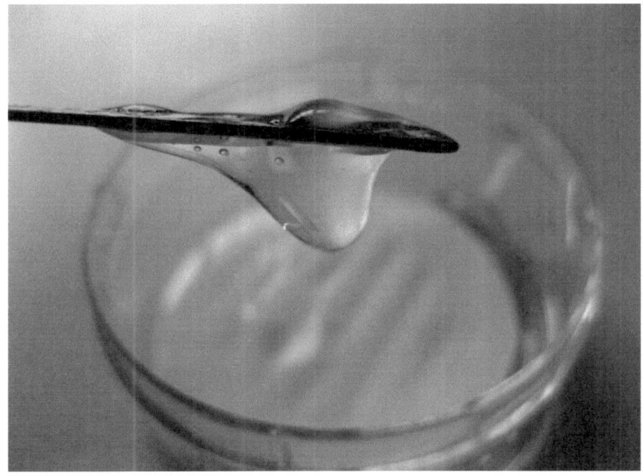

Figure 4-18: Filter cake composition of hyaluronic acid at 4 bars-4 V/mm

4.3.2 Characterization of separation of hyaluronic acid

FT-Raman spectroscopy

The alteration of the chemical structure of hyaluronic acid is another criterion to evaluate the applicability of electrofiltration. In order to detect possible changes in structure due to the applied electric field, samples which are taken before and after electrofiltration were analyzed using FT-Raman spectroscopy. The spectra of the hyaluronic acid samples derived from experiments with 4 bars-0 V/mm, 4 bars-2 V/mm, 4 bars-4 V/mm and sample of the initial hyaluronic acid dispersion are presented in Figure 4-19.

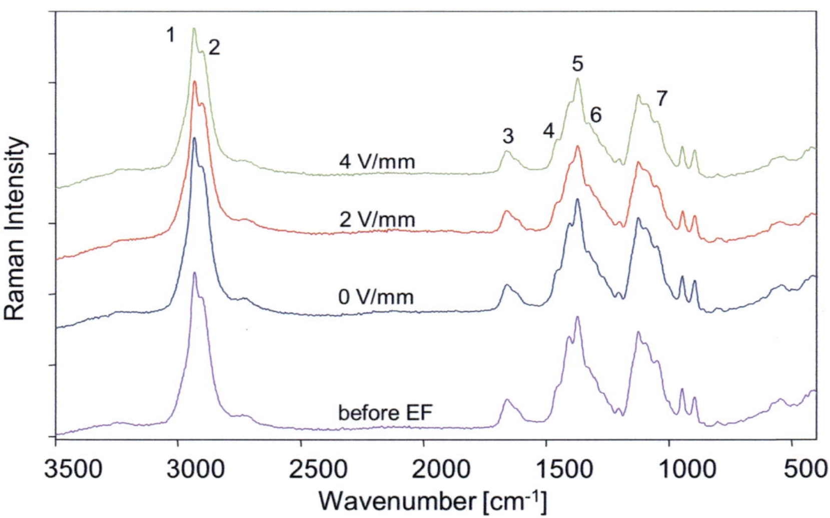

Figure 4-19: Raman spectra of HA samples before and after electrofiltration at a pressure of 4 bars and electric field strengths of 0, 2 and 4 V/mm

The main bands comprise of N-H stretching (1), C-H stretching(2), Amid I and C=C stretching (3), C-N stretching and C-H deformation (4), C-H bend (5), Amid III (6) and C-C and C-O stretchings (7) (Alkrad et al. 2003). The profile of the Raman spectra correlates with the results obtained by Alkrad et al. (2003).

The results reveal no significant difference between the spectra before and after electrofiltration and respectively the different experimental conditions. Therefore, the application of electrofiltration expresses no negative effect on the structure of HA molecules. Slightly differences between the intensity of the hyaluronic acid spectra were caused by diminutive differences in particle size. However, in general there was no detectable oxidative or reductive change of any functional group induced by electrofiltration.

Molecular weight

Another important product characteristic of hyaluronic acid influencing its properties is the molecular weight. With no electric field applied, an average molecular weight of 451 kDa was determined. After electrofiltration with an electric field of 4 V/mm average molecular weight was 462 kDa. The evaluation of possible structural changes and characteristics arising from the purification method demonstrated that electrofiltration is a relevant and promising technique in downstream processing of hyaluronic acid delivering pure product with desired unchanged specifics.

This research suggests electrofiltration as a novel downstream processing method in the production of hyaluronic acid. Limitations like process duration, energy consumption and the necessity of organic solvent use, characteristics for conventional extraction processes were overcome meeting the current downstream processing trends. The results confirmed that the applied electric field increased filtrate flux and reduced process time for hyaluronic acid dispersions. It has been demonstrated that the average molecular weight and structural composition of hyaluronic acid remained the same after introduction of electric field during processing. In addition, this scientific work explores the influence of different buffer solutions upon filtration kinetics. The obtained results established electrofiltration as an alternative technology to the extraction processes involving precipitation. For the optimization of the purification process and its successful implementation in industrial scale technologies, further experiments including hyaluronic acid obtained after biotechnological production are required.

5 Analysis of technical process and optimization

The electrokinetic and transport phenomena occurring inside the filter system were investigated by conducting several experimental series. In the first part, conductivity of buffer solutions and biopolymer dispersions was used as variable in order to analyze the effect of conductivity on filtration kinetics. In the second part of the experiments, characterization of the voltage distribution in the filter chamber was done by using the established chamber system with voltage sensors. Experiments followed by a subsequent test of different filter media to optimize the processing costs. In the third part, the designed filter chamber system with fluorescence sensors was used in the experiments for the determination of concentration gradient in the filter chamber.

5.1 Effect of conductivity

Electrofiltration experiments concerning the transition of ions in the system were performed by using model product xanthan with a concentration of 2 g/L. Owing to the stability of the biopolymer, pressure vessel was not cooled. Phosphate buffer was circulated in the system for flushing of the electrodes and a conventional filter chamber without sensors was used.

Experiments with varied conductivity in the filter chamber and flushing chambers were carried out and the influence of the conductivity on filtration kinetics was evaluated. Analysis of filtrate samples was performed by ion chromatography revealing the concentration and the variety of migrating ions. In downstream processing operations, the solutions are mostly desalinated since reduction of salinity will benefit the energy efficiency (Mehanna et al. 2010; Jacobson et al. 2011). Therefore, the analyses together with process data represent valuable information about ion exchange through the membranes which promotes optimization of the process control. The experiments were performed in two sections. First, the effect of equal conductivity and subsequently, the effect of different conductivity between filter chamber and flushing chambers upon the efficiency of the process were investigated. Conductivity of filtrate solution was measured manually during the experiments. Effect of equal conductivity on filtration kinetics was determined at low and relative high salt concentrations. In the first series of the experiments, xanthan was dispersed in distilled water without salt addition and the conductivity of the phosphate buffer was prepared equally to the xanthan dispersion (250-300 µS/cm). Different pressures and voltages were applied to the system, provided that applied parameters were kept constant during the experiments. Figure 5-1 presents the increase in filtrate mass of xanthan dispersion without salt addition for experimental series using a buffer solution with a same conductivity as xanthan dispersion at varied applied pressure and voltages.

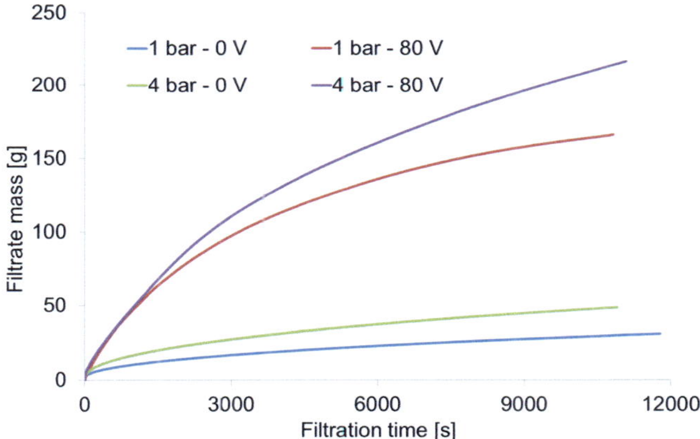

Figure 5-1: Filtrate mass-time diagram for xanthan dispersed in distilled water (~250 µS/cm), electrofiltrated using buffer solution with a conductivity of ~250 µS/cm, applied pressures of 1 bar, 4 bars and voltages of 0 V, 80 V

Figure 5-1 demonstrates that the applied pressure has a mean influence on the filtration kinetics due to the high compressibility of the xanthan filter cake. However, application of electric field increases the final amount of filtrate obviously in comparison to the experiments without electric field. Moreover, the difference in filtrate mass at 80 V between the experiments at 1 bar and 4 bars increases with time continuously whereas at 0 V a constant behavior is observed. These results correlate entirely with the filtration kinetics results of the biopolymers in chapter 4.

The effect of equality of low conductivity in filter chamber and flushing chambers was further evaluated by ion chromatography analysis of the filtrate samples together with the sample of initial buffer solution. Bar charts in Figure 5-2 present the change between the concentration of two main anions – phosphate and sulphate. The other determined ions were not in comparable levels and for that reason were not presented.

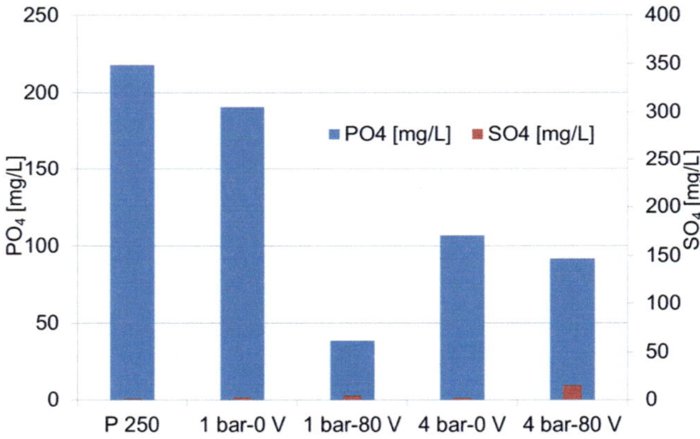

Figure 5-2: Phosphate and sulphate concentrations in the samples obtained after experimental series with equal low conductivity in chambers

The sample which was abbreviated as P250 (see Figure 5-2) is the initial phosphate buffer solution in the filtrate vessel with a conductivity of ~250 µS/cm. Considering that the buffer comprises of dipotassium-biphosphate and potassiumdiphosphate, initial buffer solution had a high phosphate concentration (218 mg/L) and lower sulphate concentration (0.93 mg/L). After performing electrofiltration at different pressures and voltages, ion concentrations in the filtrate vessel changed. The determined sulfate ions in the filtrate samples originated mainly from the filtrated xanthan dispersion and the phosphate ions originated from the buffer itself and partly from xanthan dispersion. As demonstrated in Figure 5-2, phosphate concentration in filtrate samples decreases and sulfate concentration increases in comparison to the initial values in P250 for all the experiments. This reveals that there is a considerable ion exchange through the membranes between flushing chambers and filter chamber.

Without application of electric field (at zero Volt), the results are mainly related to the dilution grade of the ions by the increase of filtrate mass which was demonstrated in Figure 5-1. Due to the applied pressure, phosphate ions in xanthan dispersion (~5 mg/L) were driven from the interior to the exterior. Increase in pressure increased the amount of filtrate and consequently decreased the concentration of phosphate ions, although some phosphate ions originating from xanthan dispersion passed from filter chamber into the flushing chambers. The stress applied on the system such as concentration difference of ions between chambers suggests a new equilibrium in the system. Related to the change of conditions, equilibrium moves to counteract the change. However, the transport of phosphate ions from the exterior (flushing chamber side of the membrane) to the interior (filter chamber side of the membrane) was insignificant since the membranes and the presented ions had both negative charge. Considering the amount of phosphate in xanthan dispersion was low, dilution affected the result dominantly. On the other hand, sulfate concentration in filtrate (for 1 bar and 4 bars) remained almost the same, as the increase of sulfate transport into the flushing chambers was balanced by the increase of filtrate. The constant conductivity of the filtrate in the absence of electric field provided no additional effect of the concentration difference through the membranes.

With application of electric field, molecules and ions start to migrate depending on their charge and the equilibrium of the ions between chambers. During experiments, pH was kept constant by buffer ions which dissociate between hydrogenphosphate, dihydrogenphosphate, phosphoric acid and phosphate. In addition, water electrolysis occurred and the formed H_3O^+ and OH^- ions react with the buffer ions (see equations 5.1, 5.2, 5.3 and 5.4).

<u>Anode</u>

$$HPO_4^{2-} + H_3O^+ \leftrightarrow H_2PO_4^- + H_2O \qquad\qquad 5.1$$

$$H_2PO_4^- + H_3O^+ \leftrightarrow H_3PO_4 + H_2O \qquad\qquad 5.2$$

<u>Cathode</u>

$$HPO_4^{2-} + OH^- \leftrightarrow PO_4^{3-} + H_2O \qquad\qquad 5.3$$

$$H_2PO_4^- + OH^- \leftrightarrow HPO_4^{2-} + H_2O \qquad\qquad 5.4$$

Figure 5-2 demonstrates that introduction of electric field into the system led to a decrease in the determined phosphate concentration in the filtrate samples at a constant applied pressure in comparison to the experiments without electric field. Additionally to the reasons explained for the experiments in the absence of electric field, under a constant voltage some other electrochemical reactions can occur beside the dissociations above (Cañizares et al. 2009). Due to the electrolysis of water and phosphate solution, water electrolysis products are produced and therefore the ionic interactions can cause the reduction of determined phosphate ions in the filtrate samples. Moreover, increase of pressure from 1 bar to 4 bars revealed that the phosphate concentration in filtrate samples at 4 bars (from 0 V to 80 V) decreased less in comparison to the experiments at 1 bar (from 0 V to 80 V). This result could be based on the reduced production of electrolysis gases at higher pressures (Weber and Stahl 2002).

Increase in pressure (at 80 V) resulted in more sulfate concentration in filtrate. This indicates that pressure and electric field together increases the release of sulfate ions from filter chamber through the flushing chambers. Furthermore, introduction of electric field resulted in a slightly decrease of conductivity in the filtrate samples due to the improvement in filtration kinetics. The occurring concentration difference assisted the transport of sulfate ions caused by pressure difference.

Further electrofiltration experiments to investigate the equality of the conductivity between filter chamber and flushing chambers were performed at higher conductivities. The conductivity of xanthan dispersion was increased to ~3000 µS/cm by using 20 mM MgSO4 solution and the conductivity of buffer solution was equaled to the conductivity of xanthan dispersion. Figure 5-3 presents the increase in filtrate mass of xanthan dispersion with salt addition for different applied pressure and voltages by using a buffer solution with same conductivity as xanthan dispersion. Figure 5-3 demonstrates that the applied pressure without electric field has no significant influence on the filtration kinetics at high conductivites as well. Application of electric field increases the final amount of filtrate obviously in comparison to the experiments without electric field. However, the comparison between Figure 5-1 and Figure 5-3 indicates that increase of conductivity decreases the reached amount of filtrate mass. The result is caused by the different zeta potentials of the xanthan dispersions with and without salt addition.

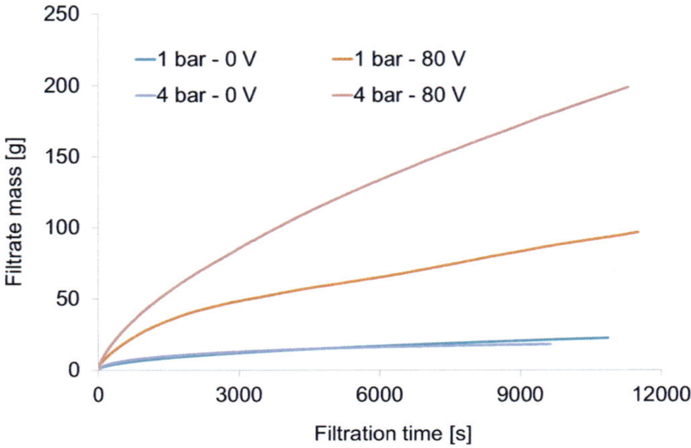

Figure 5-3: Filtrate mass-time diagram for xanthan dispersed in salt solution (~3000 µS/cm), electrofiltrated using buffer solution with a conductivity of ~3000 µS/cm, applied pressure of 1 bar, 4 bars and voltage of 0 V, 80 V

The zeta potential of xanthan with higher conductivity was -24 mV whereas the zeta potential of xanthan with lower conductivity was -40 mV at pH 5.8 (see Figure 5-4). Added salt ions countered the charge of the molecule hindering its mobility rate in electric field. In addition, this phenomenon under electric field resulted in a higher difference between filtrate masses by the increase of pressure.

The effect of equality of high conductivity between filter chamber and flushing chambers was evaluated by ion chromatography analysis of the filtrate samples together with the initial buffer solution. Figure 5-5 presents the concentration of phosphate and sulfate anions in filtrate samples and initial buffer solution with high conductivity.

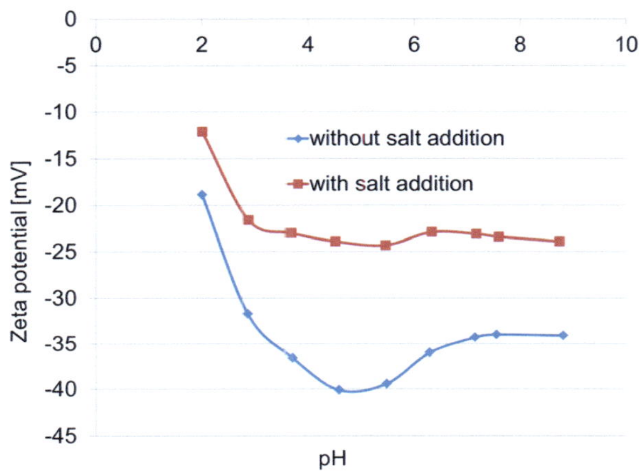

Figure 5-4: Zeta potential diagram of xanthan dispersed in distilled water (without salt addition) and dispersed in MgSO$_4$ solution (with salt addition) at different pH values

The sample which was abbreviated as P3000 is the initial buffer solution with high conductivity (3000 µS/cm). As it is demonstrated, without application of electric field by the increase of pressure, phosphate concentration remained almost same. Higher conductivity of xanthan dispersion resulted in a dense structured filter cake with lower porosity. Debye-Hückel parameter (see section 2.2) was decreased due to the high ionic concentration. As a result, lower amount of filtrate was delivered in comparison to the experimental series with equal low conductivity. Therefore, similar amount of filtrates at different pressures (without electric field) resulted in similar dilution rate of phosphate.

On the other hand, sulfate concentration increased with the increase in pressure at 0 V. The result indicates that pressure has a noticable effect at higher salt concentrations for the transition of sulfate ions through the membrane. This was based on the higher sulfate concentration in the filter chamber which consequently introduced a higher concentration difference forcing ions in the direction of flushing chamber.

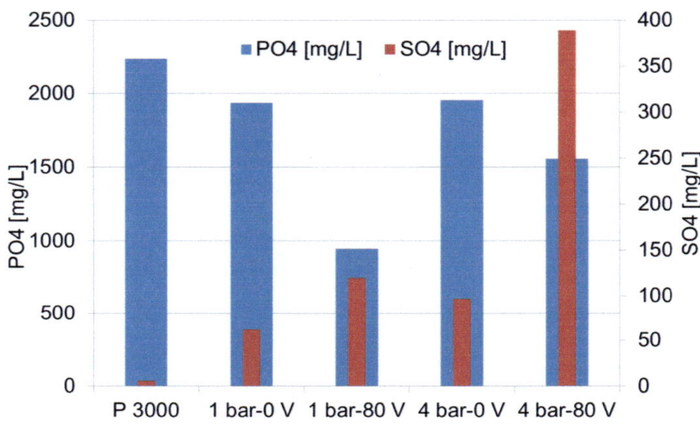

Figure 5-5: Phosphate and sulphate concentrations in the samples obtained after experimental series with equal high conductivity in chambers

Application of electric field led to an increase in the mobility of ions. The change in phosphate and sulfate concentration by application of electric field was due to the migration of ions in the system and the effect of electrolysis products as discussed in previous experimental series. With higher sulfate concentration in xanthan dispersion, the release of sulfate ions by application of electric field was achieved in higher amounts. High sulfate concentration difference between the chambers induced the increased transport of the ions together with the effect of pressure and electric field. The high ion permeation capacity in high conductivity medium can be valuable and applicable when working with dispersions from cultivation medium.

Conductivity measurements revealed the similar tendency as in the first experimental series. Without electric field, conductivity remained almost same revealing that the increase of filtrate mass compensated the increase of conductivity by transported ions. The application of electric field caused slightly decrease in conductivity in the filtrate vessel. This demonstrates that the increase of filtrate mass was high and led the conductivity to decrease despite of permeated ions.

Further experiments were performed to investigate the inequality of the conductivity between filter chamber and flushing chambers. The conductivity of xanthan dispersion was ~250 µS/cm (without salt addition) and the buffer solution was prepared with a conductivity of ~3000 µS/cm. Figure 5-6 presents the increase in filtrate mass of xanthan dispersion without salt addition for different applied pressure and voltages by using a buffer solution with high conductivity.

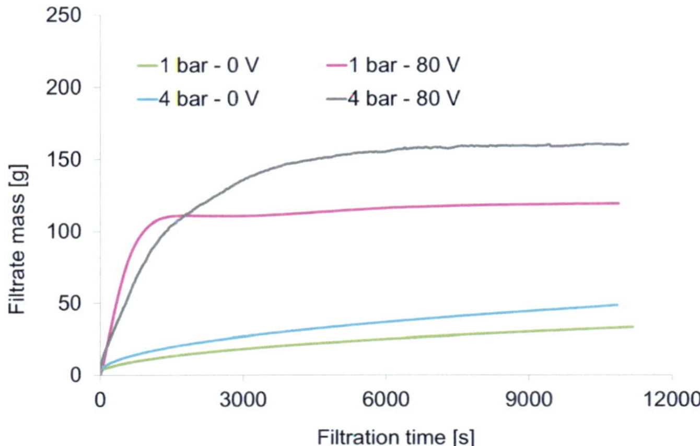

Figure 5-6: Filtrate mass-time diagrams for xanthan dispersed without salt addition (~250 µS/cm), electrofiltrated using buffer solution with a conductivity of ~3000 µS/cm, applied pressure of 1 bar, 4 bars and voltage of 0 V, 80 V

Increasing of applied pressure without electric field had again very slightly effect on filtration kinetics of xanthan dispersion. However, application of electric field with pressure presented interesting results. Differently than the previous series with equal low and high conductivity in chambers, filtrate mass did not increase continuously in the experimental time. When xanthan dispersion was electrofiltrated using concentrated buffer solution, filtrate mass remained constant in a short time after starting experiment. The result indicates the high effect of the ionic changes when conditions inside and outside the filter chamber are not equal. The equilibrium required to move to counteract the change in the direction of filter chamber. However, due to the applied pressure and the charge of the membranes, diffusion was not possible from the exterior to the interior. As a result, membrane potential on the surface changed and the high ionic strength of buffer solution opposed the transition of the ions intensively from filter chamber into the flushing chamber.

Disequilibrium of conductivity in the chambers decreased the passing capacity of filtrate through the membranes. The required time to reach the constant state was greater at 4 bars-80 V. This result was based on the dominative pressure effect despite the inhibiting effect of the disequilibrium. Confirmation of the results was achieved in details by ion chromatography analysis which was presented in Figure 5-7. The figure presents the change in phosphate and sulfate ions after electrofiltration experiments when using xanthan dispersion with low conductivity (without salt addition) and buffer solution with high conductivity. Differently from the previous series, phosphate bars follow a stepwise concentration decrease by the variation in conditions. Due to the slightly increase of the filtrate mass without application of electric field, phosphate concentration decreased with the increase of pressure. On the other hand, sulfate concentration revealed almost no improvement by the increase of pressure without electric field. The low initial sulfate amount in the xanthan dispersion and consequently a low sulfate concentration difference between the interior and the exterior caused the result as in the first experimental series with equal low conductivity in the chambers.

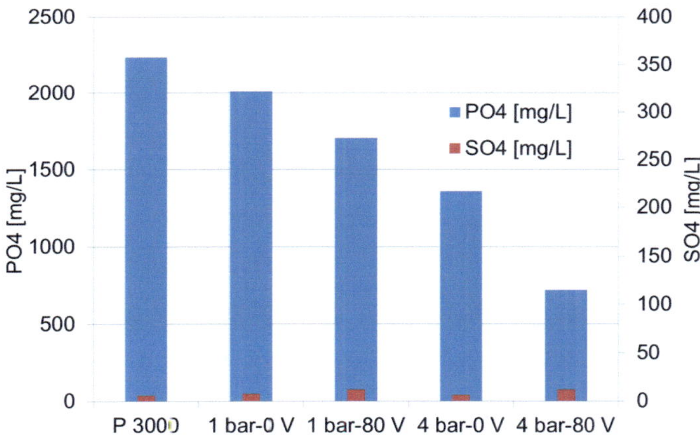

Figure 5-7: Phosphate and sulphate concentrations in the samples obtained after experimental series with low conductivity in the filter chamber and high conductivity in the flushing chambers

With application of electric field at 1 bar, phosphate concentration did not decrease as much as in the other high conductivity buffer series compared with the results at 1 bar without electric field. Concerning the high phosphate concentration difference, the ions accumulated on the exterior side of the membrane surface resulting in a constant behavior of the filtrate mass under a pressure of 1 bar after 15 min. Due to the accumulation effect, the phosphate ions remained in the filtrate solution relatively in higher amounts. Beside this effect, at 4 bars with electric field, phosphate concentration decreased more presumably due to the increased filtrate mass, less production of electrolysis gases and the more filtration time till the pores were blocked. Manual conductivity measurements during experiments verified the obtained results. Without application of electric field conductivity did not change substantially. However, with application of electric field, the conductivity decreased at the beginning of the experiments drastically revealing the fast increase in filtrate mass (see appendix A1). After reaching the constant state, the conductivity values did not change vastly. On the other hand, sulfate concentration in the filtrate samples increased in comparison to the cases without electric field. However, increasing applied pressure under electric field resulted in no improvement in the sulfate concentration. High forcing effect of phosphate concentration difference from the exterior to the interior could increase the membrane potential and therefore opposed the transport of sulfate ions into the flushing chamber partially. In these experimental series, the results were affected strongly by the constant behavior of the filtration kinetics resulting from the change of membrane potentials.

The inequality of the conductivity between filter chamber and flushing chambers was further investigated by using xanthan dispersion with relative high conductivity (~3000 µS/cm) and a buffer solution with lower conductivity (~250 µS/cm). Figure 5-8 presents the increase in filtrate mass of xanthan dispersion with salt addition for different applied pressure and voltages by using a buffer solution with lower conductivity.

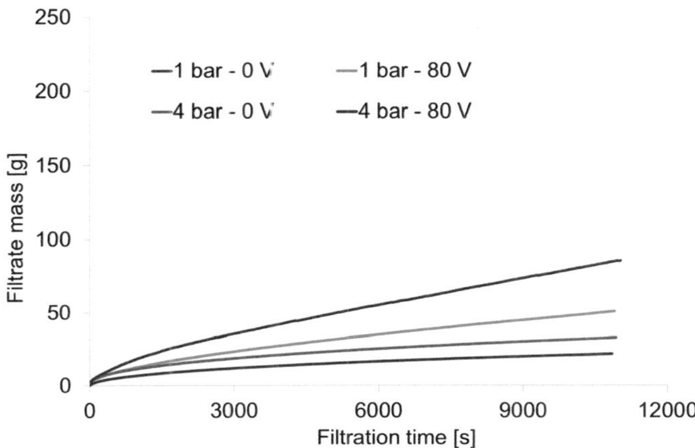

Figure 5-8: Filtrate mass-time diagram for xanthan dispersed with salt addition (~3000 µS/cm), electrofiltrated using buffer solution with a conductivity of ~250 µS/cm, applied pressure of 1 bar, 4 bars and voltage of 0 V, 80 V

The experiments with high interior conductivity and low exterior conductivity demonstrate that application of electric field under pressure has a mean effect on filtration kinetics. Although the xanthan dispersion was prepared identically (with salt addition) to the one used for experimental series with high equal conductivity in the chambers, the results reveal different filtration behaviors. The reason arises from the potential ionic migrations and the interactions through the membranes. Therefore, the experiments reveal the importance of the conductivity difference between the interior and the exterior. Without application of electric field, these effects are negligible since the buffer concentration in this case has no significant influence. Using identical xanthan dispersion results in similar cake properties and subsequently similar filtration behaviors. Therefore, the obtained amounts of filtrate under no electric field were close to the values for the series with high equal conductivity in the chambers.

The discussion of the results can be further explained by the conductivity and pH measurements. Differently than the other series, conductivity in the filtrate samples increased for all the experiments, since the filtrated dispersion has higher conductivity in comparison to the buffer conductivity (see appendix A2). Moreover, conductivity increased intensively by application of electric field compared with the experiments without electric field. Migration in electric field increased the transition of salt ions with the filtrate. In addition, pH decreased only for the experiments with electric field applied. This demonstrates that the buffer cannot support the system anymore to keep the pH stable due to the ionic change through the membranes. Figure 5-9 presents the concentration of initial buffer solution with low conductivity and concentration of phosphate and sulfate anions in filtrate samples after the experiments with high interior conductivity and low exterior conductivity. As demonstrated in Figure 5-9, transition of sulfate ions remained same without application of electric field. However, by application of electric field the concentration of sulfate ions in the filtrate samples increased, due to the high sulfate concentration difference and the migration of ions in the electric field. On the other hand, increasing pressure under electric field revealed higher transition of sulfate ions from the interior to the exterior. Concentration of phosphate ions decreased with application of electric field. Nevertheless, at 4 bars-80 V, phosphate concentration remained same in comparison to the experiment without electric field. The reason could be based on the high sulfate concentration as phosphate concentration. The unstable conditions arising from the ionic interactions in the filtrate could give this result.

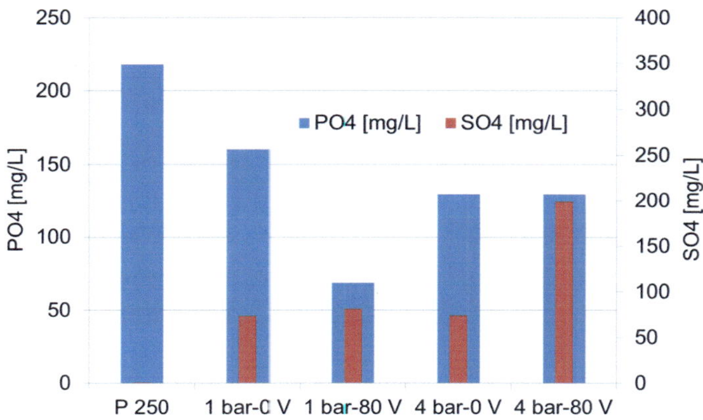

Figure 5-9: Phosphate and sulphate concentrations in the samples obtained after experimental series with high conductivity in the filter chamber and low conductivity in the flushing chambers

Briefly, it is obvious from the performed experiments that the ion exchange through the membranes plays an important role. Hence, the concentration of the buffer and choice of buffer (see chapter 4) are important factors affecting the purification process. The experimental series with different interior and exterior conductivities reveals that electrofiltration can be applied successfully for the biodispersions with high conductivity direct after cultivation. High transition capacity of the ions through the membranes purifies the dispersion from unwanted impurities. On the other hand, the higher the conductivity is, the higher the current for the processing. However, even at high conductivities the energy needed is lower in comparison to the conventional methods for the separation and purification of biopolymer dispersions. In addition, by

132

using a buffer change system during electrofiltration, the conductivity in the filtrate vessel can be decreased and subsequently the current needed (see section 6.1). Besides, experiments demonstrated that the most efficient results could be received when the buffer and biopolymer dispersion conductivities were not extremely different.

5.2 *In-situ* voltage characterization

The voltage drop inside the filter chamber is critically important when electric field is applied between anode and cathode electrodes. Since there are several components in the electrofiltration system that can increase the total resistance, it is necessary to know the voltage distribution through the system for further optimization of the filter chamber and whole system design.

As illustrated in section 3.4.3, the system comprises of three components (two flushing chambers and a central filter chamber) with a cross-sectional length of 20 mm. In order to analyze the voltage distribution through the chambers, experiments were performed with the designed filter chamber constructed with voltage sensors (see Figure 3-6). The filter chamber was divided into four sections by five of the sensors. The rest two sensors were placed with 1 mm distance from electrode surfaces. The Figure 3-8 demonstrates that effective voltage consists of the sum of U_1, U_2, U_3 and U_4. Each of these measuring points has 2 mm distance to the next voltage sensor. However, U_A and U_C were detected in a larger distance (5 mm each) and include the flushing chamber sections. In other words, U_A and U_C are the voltage drops between the first (close to electrode surface) and the second sensors (close to the interior membrane surface) from anode and cathode sides, respectively. The filter components such as membranes, supporters and the grids are inclusive to these parts on both sides. U_{aa} and U_{cc} values were measured in 1 mm distances on both sides between the electrodes and the first sensors.

The experiments were conducted using xanthan dispersion with a concentration of 2 g/L. For buffer solution, phosphate buffer with various conductivities were used during the experiments. All experiments were performed applying moderate chosen conditions – 1 bar pressure and 40 V voltages. Figure 5-10 demonstrates the progression of voltage in the electrofiltration system for an experimental time. U_{aa} and U_{cc} were measured manually and presented as sum of them.

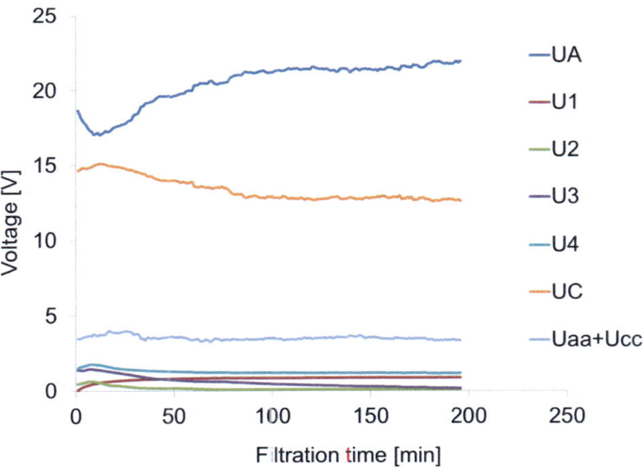

Figure 5-10: Voltage progression in the system

Analysis of the distance between sensors designates that U_A, U_C, U_1, U_2, U_3, U_4, U_{aa} and U_{cc} acquires 25%, 25%, 10%, 10%, 10%, 10%, 5% and 5% of the total applied voltage respectively. These percentages correspond to 10 V, 10 V, 4 V, 4 V, 4 V, 4 V, 2 V and 2 V respectively for an experiment with an introduced voltage of 40 V. Therefore, the effective voltage in the filter chamber (U_{eff}) corresponds to 40% (16 V) of the applied voltage. However, the distribution of voltage through the system does not fit in with the theoretical expectations. The voltage drop mostly stands for U_A and U_C. The result was affected by the membranes, supporters and the grids which were placed in these sections. Each component has an additional resistance that decreases the access of the current.

Moreover, during the experiment, current progression of anode and cathode detections demonstrated a mirror-inverted depiction. This characteristic can be explained by concentration increase resulting with a cake composition on anode side membrane, due to the negatively charge of xanthan molecules. The accumulation of the biopolymer on anode side membrane increased the voltage drop on this side. On the other hand, cathode side was almost free of filter cake and led filtrate to flow resulting in decrease in current intensity. The increase in U_A and U_C caused a decrease in effective voltage (U_{eff}) in the filter chamber. Figure 5-11 represents the small changes of U_1, U_2, U_3 and U_4 for a better visualization of the interior voltage drop. Excluding U_1, all the other sections inside the filter chamber demonstrated a decreasing behavior in voltage compared with the beginning of the detections. U_2 and U_4 remained constant after an hour, revealing that the resistance in the system demonstrated a constant pathway. However, U_1 increased from the beginning of the measurements until the resistance reached a constant value. This was based on the cake accumulation as happened for the U_A sensors. The reproducibility of the results was confirmed by the several repetitions of the experiments.

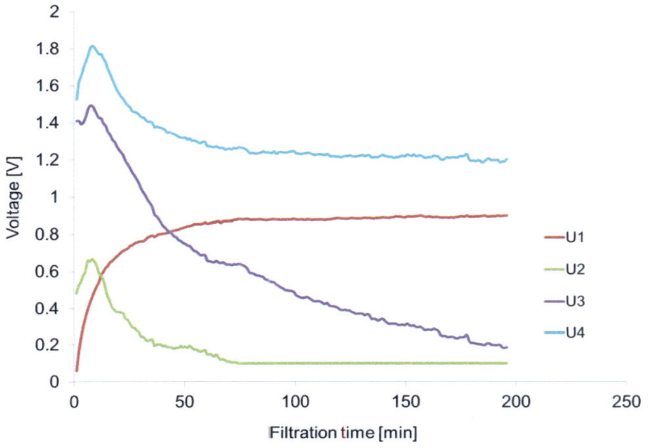

Figure 5-11: Voltage progression in the filter chamber

Since current remained almost the same during the experiments, an increase in voltage designates an increase in the resistance as well. Moreover, U_2, U_3 and U_4 revealed an increase for a short time at the beginning of the experiment resulting from the hydrodynamic resistance force which was explained in chapter 1. U_3 decreased continuously differently than the curves of U_2 and U_4. Position of the section U_3 corresponds to the entrance of the biopolymer dispersion. Therefore, the continuous decrease in U_3 was affected by the biopolymer flow into the filter chamber. The results confirm that filter cake thickness contacted to the sections U_A and U_1 properly.

Determinations reveal that the sum of anode and cathode side voltage drop reaches up to 84% and the effective voltage up to 7%. However, the sum of voltage drop of U_A and U_C must have been around 50% theoretically. The increase in voltage drops of U_A and U_C decreased the access of the current through the filter chamber and consequently resulted in a lower interior voltage. The voltage distribution was demonstrated in percentages on Figure 5-12 separately for each measuring zones.

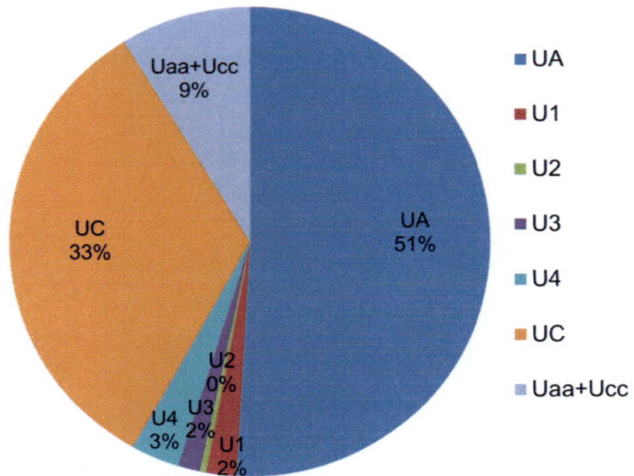

Figure 5-12: Voltage distribution in the whole electrofiltration system

The voltage drop on anode side (U_A) was much more than cathode side voltage drop (U_C). This demonstrates that the filter cake accumulation on anode side membrane affects the voltage drop distribution in addition to the filter components which increase the resistance. As a result, the effective voltage inside the filter chamber remains around 3 V. The limitations which have curtailed the use of total applied voltage in the chamber system have to be identified for further improvements.

Based on the results presented in section 5.1, experiments were performed aiming to comprehend the transport phenomena and their internal effects on filtration kinetics. Xanthan dispersion was prepared without salt addition and the electrodes were flushed by phosphate buffer solutions with various adjusted conductivities. Filtrate mass dependent progression of effective voltage inside the filter chamber was presented in Figure 5-13.

The effective voltage followed a decreasing tendency 20 minutes after starting the experiment for all the experiments flushed using buffer solutions with different conductivities. However, increasing the conductivity of flushing solution increased the starting effective voltage. The higher the conductivity, the higher was the decreasing range of interior voltage drop. This was related to the current progression of these experiments (see appendix A3). Higher conductivity resulted in a higher current value in the system proportional to the applied conductivity. Consequently, more ions were accumulated in membrane pores and a higher interior voltage was achieved. With passing time, filter cake accumulation increased and filter cake resistance increased as well. Therefore, the effective voltage could not keep the high value and decreased continuously down to the range 2.5-3.8 V. In addition, average voltage distribution revealed no significant change by using buffer solutions with different conductivities. U_A, U_C and U_{eff} values ranged between 51-54%, 28-35 % and 5-8% respectively. However, the results highlight the phenomenon which was discussed in the previous section.

In the section 5.1, it had been clarified that using very high concentrated buffer solutions blocks the filtration and subsequently filtration kinetic reaches a constant state faster in comparison to the experiments conducted by using buffer solutions with equal conductivity as xanthan dispersion's one. The experiments performed with the filter chamber designed with voltage sensors revealed that conductivity of the flushing solutions can be increased until a specific value. After a critical conductivity point, filtration is blocked. However, this critical point can be product related and should be determined for each product separately for industrial applications.

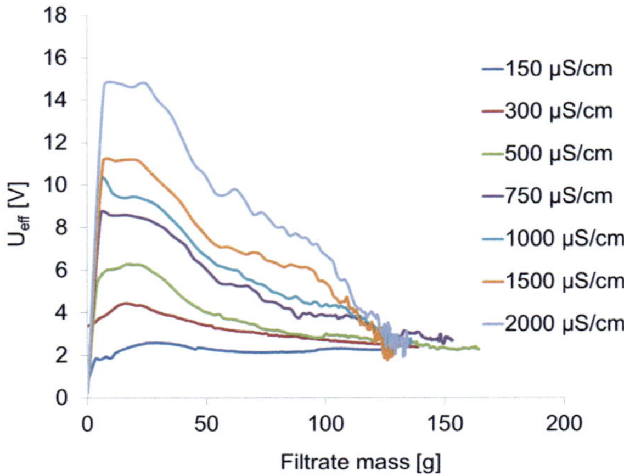

Figure 5-13: Voltage drop inside the filter chamber when flushed with buffer solutions with various conductivities

Filtrate mass versa time diagrams of these experiments were presented in Figure 5-14. The experiment performed using a buffer solution with a conductivity of 500 µS/cm presents the highest mass of filtrate with an increasing tendency. Therefore, the conductivity of buffer solution can be prepared up to twice higher than the conductivity of xanthan dispersion. Flushing the electrodes with a buffer solution which has a lower conductivity than the conductivity of product dispersion, resulted in lower filtrate mass compared with the experiments performed by using equal conductivities. Especially, too low conductivities, e.g. 8 µS/cm demonstrated remarkably low filtrate mass. Therefore, electrodes should not be flushed with distilled water or buffer solutions with too low conductivity due to the absence of current passage. Increasing the buffer conductivity contributed the filtration kinetics until a specific value. However, increasing buffer conductivity up to 750 µS/cm led to a decrease in filtrate mass compared with the experiment performed by using a buffer solution with a conductivity of 500 µS/cm. In addition, application of a buffer with 1000 µS/cm conductivity and more revealed a constant filtration behavior after almost an hour. The results were caused by the blocking effect which was discussed in section 5.1.

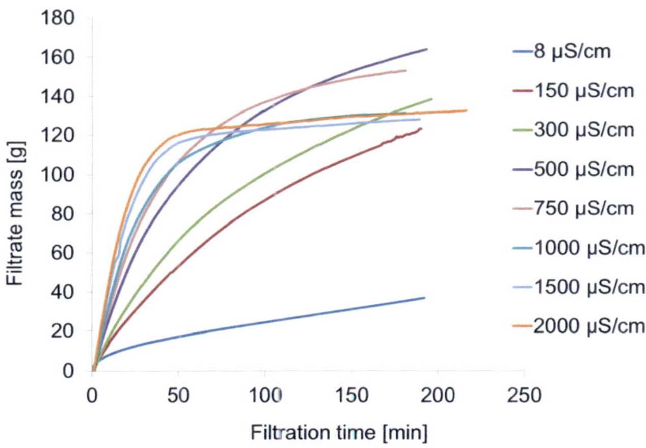

Figure 5-14: Filtrate mass versa time diagram of the experiments using various adjusted flushing solution conductivities

On the other hand, in order to test the applicability of the obtained result, interior voltages and the energy consumptions were compared. Energy consumption is not only related to the current intensity which is proportional to the conductivity of the solution but also to the filtrate mass. The calculations revealed that the experiment performed by using the buffer solution with a conductivity of 500 µS/cm needed 364 J less energy per 100 g filtrate. Increasing the conductivity more than this value resulted in higher energy consumption. Therefore, insignificant and small changes in energy calculations make the application of considerably higher buffer conductivities more applicable.

Figure 5-15 compares interior voltage in the filter chamber when using buffer solutions with different conductivities under conditions of 1 bar and 40 V. Application of the buffer solution with a conductivity of 500 µS/cm demonstrated almost equal effective voltage compared with the result obtained from the experiment by application of 300 µS/cm (nearly to the xanthan dispersion's one). Increasing conductivity up to 1000 µS/cm revealed the highest interior voltage. However, due to the blockage of the filtration it was not preferrable. Moreover, Figure 5-15 designates that choosing a buffer solution wit a conductivity up to 750 µS/cm could come into consideration depending on the product's overall characteristics.

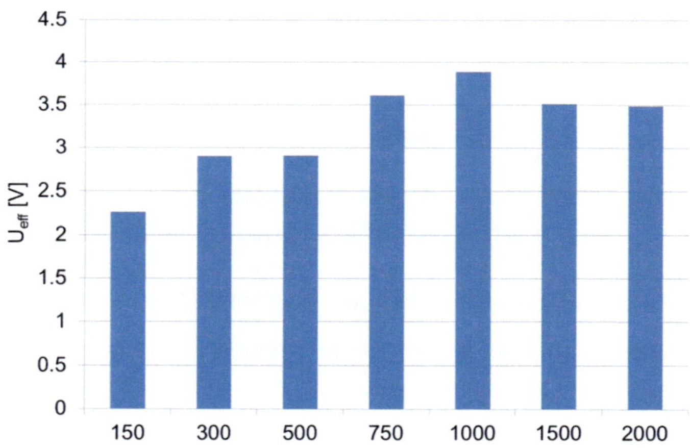

Figure 5-15: Effective voltage in the filter chamber by using buffer solutions with different conductivities

In conclusion, choosing the conductivity of buffer solution have to be decided by considering the product related filtration behavior. The inequality of conductivities between interior and exterior of the filter chamber can be taken into consideration unless the range is not too high. The specific conductivity value has to be determined by several experimental tests.

Further experiments were performed in order to identify the voltage drops in the defined sections when using different filter components and conditions. All the experiments were conducted under 1 bar overpressure and 40 V voltages with a filtration length of 1.5 hours. Phosphate buffer was used as in previous experiments. Figure 5-16 presents the differences in average voltage distribution in the chamber system with conditions with and without membranes and grids when no xanthan dispersion was used.

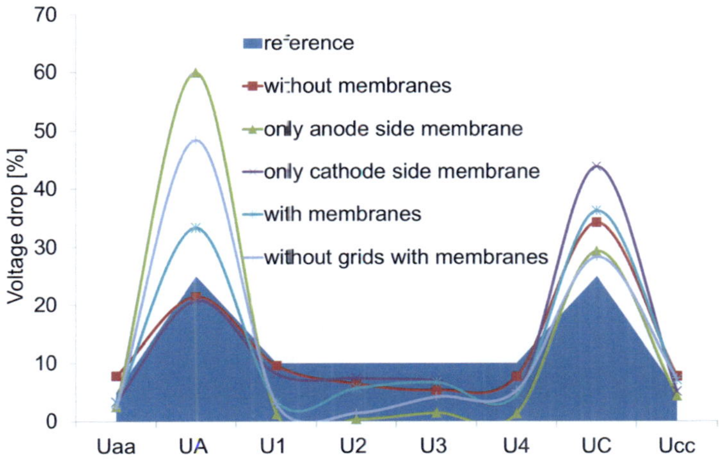

Figure 5-16: Average voltage drop distribution in the system when using with and without membranes (no xanthan dispersion)

The blue background demonstrates the theoretically estimated voltage drops considering the distances between two neighbor sensors. Additional curves reveal the deviation of the voltage drop in these sections in comparison to the reference values. Without using membranes (without supporters and with grids), anode side voltage drop followed a near value to the reference. However, from beginning U_1, it started to decrease in the filter chamber until the fourth measuring zone (U_4). After that point, the voltage drop on cathode side increased and exceeded the prescribed value. The experiment represents the effect of grids on the voltage distribution in the system. Experiments which were performed by using partly framed filter chamber (for anode and cathode sides separately) revealed that the voltage drop increased on the sides where membranes were used. However, application of only anode side membrane resulted in a higher voltage drop (up to 60%) in comparison to the case only cathode side membrane was used. This phenomenon can be presumably explained by the charge of buffer solutions. When only anode side membrane was used, ions that migrated through the system were in the same direction as pressure induced. When only cathode side membrane was used, the direction of pressure assisted flow and ionic migration was in opposite directions. As a result, the voltage drop on anode side membrane was higher compared with the voltage drop on cathode side membrane. However, it must be considered that the application of pressure denatures the form of membranes. Therefore, the membrane used experiments were performed using supporters additionally to the membranes and the achieved curves are also as a result of the resistance of supporters.

When both membranes were used the voltage drops decreased in comparison to the experiments when only one side membrane was used. Nevertheless, the voltage drops increased in comparison to the experiments when no membranes were used. This result is partially caused by the supporters. On the other hand, when no grids were placed on the electrodes, the experiments revealed less interior voltage in the filter chamber with a higher U_A and lower U_C than the experiment with all the components (named "with membranes" in the diagram). Excluding the last experiment – without grids, all the experiments presented in Figure 5-16 were conducted by using grids. This comparison demonstrates the effect of membranes and supporters.

The following figure presents the average voltage drop distribution in the system with and without xanthan dispersion. The voltage distribution curves in Figure 5-17 point the effect of filter cake accumulation on anode side membrane. Due to the negative charge of the xanthan molecules, filter cake was formed on anode side membrane. As a result, anode side voltage drop (U_A) increased from ~33% to ~50% compared with the same experimental conditions without using xanthan dispersion.

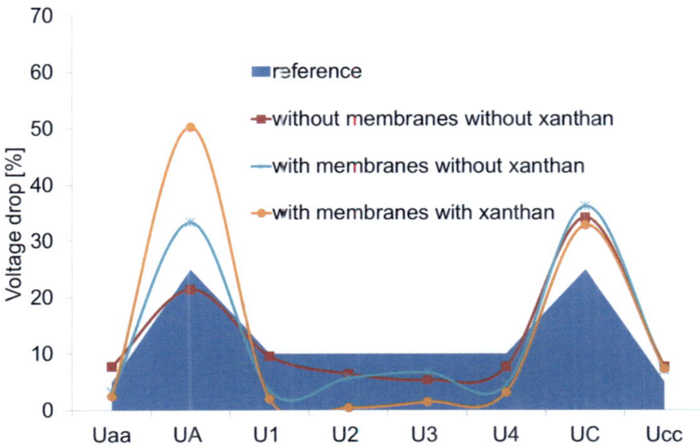

Figure 5-17: Average voltage drop distribution in the system when using with and without membranes and xanthan dispersion

These results confirm that most of the voltage drop in the system is caused by the concentration effects. Therefore, the filtration time to determine the voltage drop is another important factor. With the increase of the thickness of the filter cake, the voltage drop on anode side increased as illustrated in Figure 5-10. Consequently, only low interior voltages were achieved. On the other hand, cathode side measurements revealed almost same values since on cathode side membrane no filter cake was formed. U_{aa} and U_{cc} values remained in the same range due to the absence of any change in flushing chamber section.

Average effective voltages in percentages were ~28%, ~20% and ~7% for the experiments without membranes-without xanthan, with membranes-without xanthan and with membranes-with xanthan, respectively. The results demonstrate that in addition to the filter cake effect which reduced the effective voltage by 13% in the experimental time, the supporters and the membranes increased the total resistance and reduced the interior voltage around 8%. Further experiments can be performed in order to analyze the effect of supporter and membranes separately. Changing the form and/or material of membranes and supporters can assist to increase the effective voltage.

5.3   Membrane selection

Membranes are one of the important structural components of the electrofiltration system. Additionally to operating conditions, membrane selection can affect the filtration characteristics. Membrane materials for electrofiltration have to acquire some properties such as high chemical, mechanic and thermal stability, high porosity and low cost. The process can be optimized with respect to membrane selection. In order to test the appropriate membrane material, experiments using xanthan dispersion were performed at moderate conditions, 1 bar-40 V. Figure

5-18 presents the needed energy for 100 g filtrate when using various microfiltration membrane types.

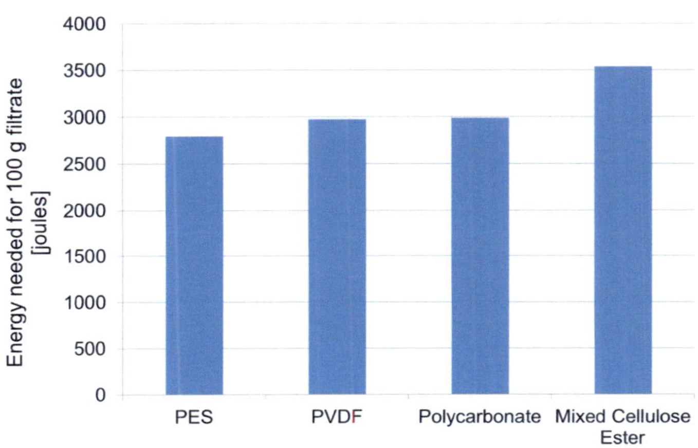

Figure 5-18: Energy consumption per 100 g filtrate for various membrane materials with a pore size distribution of 0.1 µm

Due to the capability of high throughputs, the membranes with a hydrophilic character were used in the experiments. Figure 5-18 reflects the similarity in energy consumption values of PES (Polyethersulfone), PVDF (polyvinylidene fluorid) and polycarbonate membranes. Energy consumption for PES membrane was slightly lower than PVDF and polycarbonate membranes. In comparison to those three membranes, mixed cellulose ester resulted in a higher energy need for the same amount of filtrate.

The results are related to the characteristics of the membranes such as chemical composition, porosity and flow rate capacity. For a better evaluation t_f/V_f versus V_f diagrams were presented with various membrane materials in Figure 5-19. PES membrane demonstrated less specific filter cake resistance for the same operating conditions in comparison to the other membranes. The result could be explained by the high porosity distribution (app. 74%). The membrane made of mixed cellulose ester presented the highest resistance compared with the other materials.

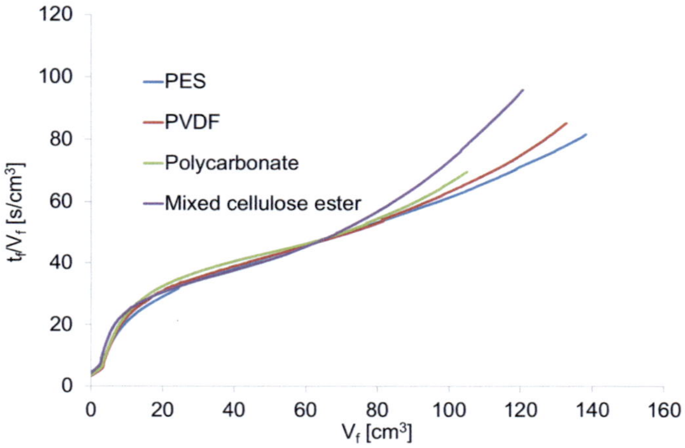

Figure 5-19: t_f/V_f versus V_f diagram of xanthan dispersions at 1 bar – 40 V when using various membrane materials

In order to investigate the current permeability of the membranes, several experiments were performed. However, the determined effective voltage (U_{eff}) remained in the range of 2.4-3.3 V for the tested materials.

Highly asymmetrical morphology of PES membrane (see Figure 5-20) offers high filtration performance with high flow rates. However, despite the attractive advantages, chemical cleaning of membrane affects the performance and surface properties of PES membrane (Arkhangelsky et al. 2007) and must be considered for the scale-up of the process. In order to increase the ion throughout, the membranes used during electrofiltration experiments were introduced a pretreatment before starting the experiments.

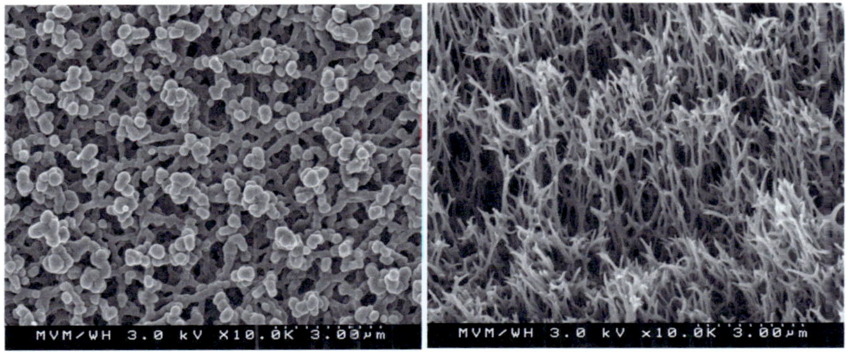

Figure 5-20: The SEM images of surface (left) and cross section (right) of the PES membrane

The both anode and cathode side membranes were immersed in the buffer solution which would be used in the experiment. When the pores of the membrane reached saturation, high ion fluxes could be achieved at the beginning of the experiment.

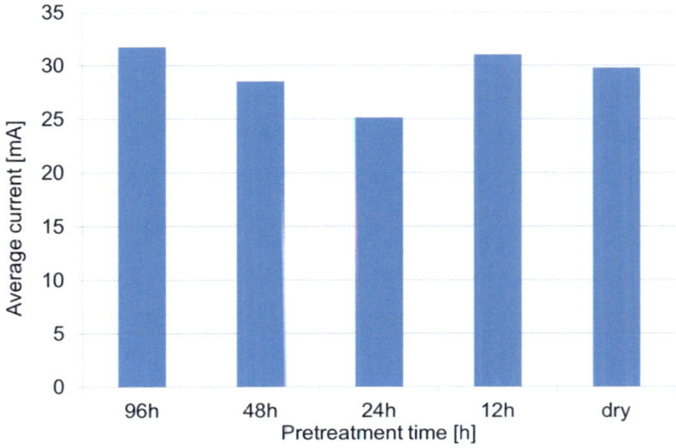

Figure 5-21: Average current intensity after experiments at 1 bar-80 V for various membrane pretreatment times

Determination of pretreatment time was performed using PES membranes. For these series the maximum available voltage (80 V) was applied. The average current values after applying different pretreatment times were presented in Figure 5-21.

After immersing the membranes in buffer solution for 24 h, the average current revealed the lowest value compared with the other pretreatment times. Since the low current corresponds to low energy consumption for the same conditions, PES membranes have to be immersed in buffer solution at room temperature one day before starting the experiment. The more or less pretreatment time resulted in higher average current due to the ionic effects in the membrane pores.

5.4 Determination of concentration gradient in the filter chamber by using fluorescence sensors

Electrofiltration experiments aiming to evaluate the concentration gradient in the filter chamber were performed by using the new designed chamber with fluorescence sensors. LED-induced fluorescence device enabled monitoring of the migration of the fluorophore labeled biopolymer molecules providing visualization of the concentration gradient. Fluorescein isothiocyanate (FITC) conjugated dextran was used to image the development.

The homogeneous mixture of FITC-dextran (av. Mol wt: 20000) and technical dextran (av. Mol. Wt. 5000000-40000000) dispersions were fresh prepared for each experiment. Labeled and unlabeled biopolymer mixture was preferred in order to minimize any perturbation effect of the

FITC moieties. 25 mg of FITC-dextran was solved in 5 mL distilled water and mixed with 1 g of dextran solved in 200 mL distilled water. In order to obtain a faster accumulation of the filter cake, maximum available electric voltage was introduced to the system.

Optical fibers sent excitation energy and emitted energy were collected by the fluorescence sensors which were connected to the spectrometer. LED-light was induced at 460 nm and fluorescence was emitted at 520 nm. Under applied conditions of 1 bar and 80 V, the fluorescence spectra of the labeled biopolymer detected by 5 sensors are presented in Figure 5-22, Figure 5-23, Figure 5-24, Figure 5-25 and Figure 5-26 individually. The sensors were named from the cathode side to the anode side (see Figure 3-11). The measurement numbers (from 1 to 6) in the figures signify the defined time points – 0, 20, 100, 180, 255 and 365 min.

Due to the electrophoretic migration of the biopolymers towards anode direction, a concentration gradient occurred in the experimental time. The measurements taken at determined time points revealed the characteristic feature of the cake accumulation in the filter chamber. The reflection (at 460 nm) and emission (at 520 nm) spectra of FITC-dextran solution which were detected by the sensor 1 at 6 different time points were presented in Figure 5-22.

Sensor 1, being the first sensor on cathode side, detected a decrease in the emission intensity after the first determination. This indicates that the concentration of labeled dextran on cathode side membrane decreases with time due to the migration of molecules towards anode direction.

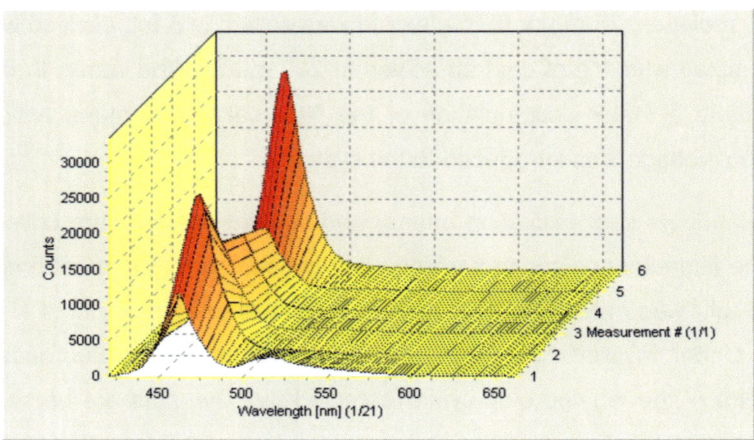

Figure 5-22: Fluorescence spectra of FITC-dextran detected by sensor 1 for the measurements at time points 0, 20, 100, 180, 255, 365 min

The emission peak after conduction of the biopolymer dispersion into the chamber (t=0) was based on the equal distribution of the concentration through the filter chamber at initial conditions. Beginning from 20 min the emission peak reached the minimal intensity confirming that no more FITC-dextran was accumulated in the first section of the filter chamber. This phenomenon can be observed further for sensor 2 in Figure 5-23. However, it took more time until the fluorescence peak of the second sensor reached essentially zero. After 180 min, no more emission of fluorescence was monitored. Fluorescence spectra of FITC-dextran obtained by sensor 3 revealed a slightly increase in intensity of peaks at 520 nm (see Figure 5-24). In the third section of the filter chamber, at which the fluorescence intensity was detected by sensor 3, a moderate concentration of the labeled biopolymer was determined during the experiment. The spectra demonstrate that the filter cake thickness approaches to this part of the chamber. In addition, concentration gradient was partly affected by the location of the sensor 3 which was close to the entrance of the filter chamber.

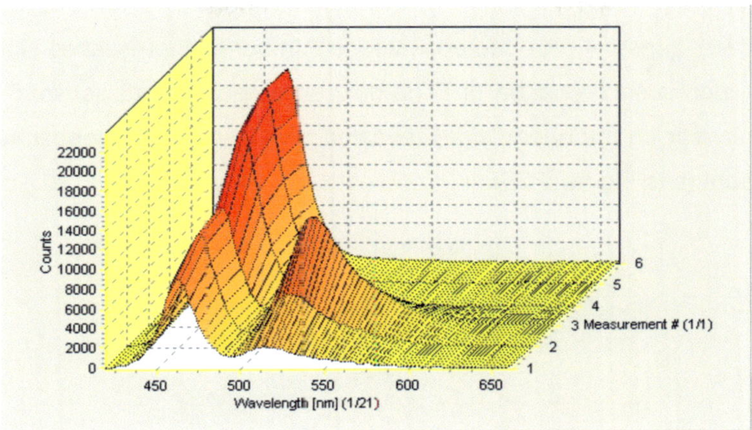

Figure 5-23: Fluorescence spectra of FITC-dextran detected by sensor 2 for the measurements at time points 0, 20, 100, 180, 255, 365 min

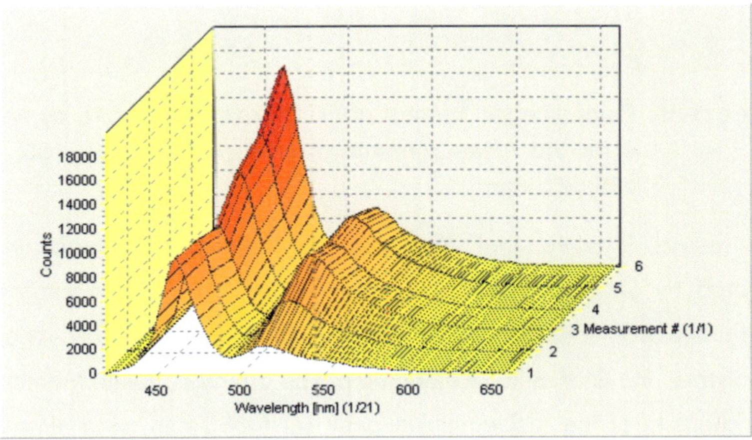

Figure 5-24: Fluorescence spectra of FITC-dextran detected by sensor 3 for the measurements at time points 0, 20, 100, 180, 255, 365 min

As demonstrated in Figure 5-25, concentration of FITC-dextran increased continuously in the fourth section of the filter chamber. Due to the electrophoretic migration of the labeled molecules towards anode direction, sensor 4 provided an increase number of fluorescence signals. The fluorescence spectra of FITC-dextran detected from sensor 5 (the first sensor on the anode side) revealed a well-structured concentration gradient (see Figure 5-26).

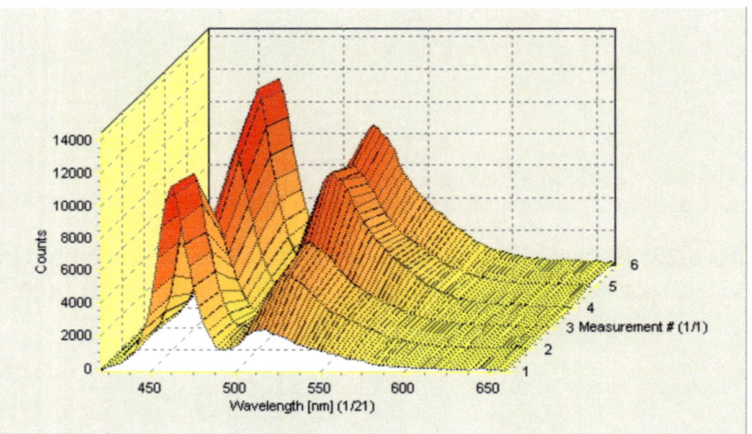

Figure 5-25: Fluorescence spectra of FITC-dextran detected by sensor 4 for the measurements at time points 0, 20, 100, 180, 255, 365 min

The results indicate that during electrofiltration the concentration detected by sensors 4 and 5 increases simultaneously starting from anode towards cathode side. However, for a positively charged biopolymer, the filter cake formation must be on the opposite membrane. The intensity of the last emission peak (at t=365 min) was almost 10 times more in comparison to the first emission peak (at t=0). This result correlates successfully with the final concentration of the electrofiltrated product which was concentrated up to 10-fold as well. 3-D visualization of concentration gradient in the filter chamber was presented in Figure 5-27 at various time points.

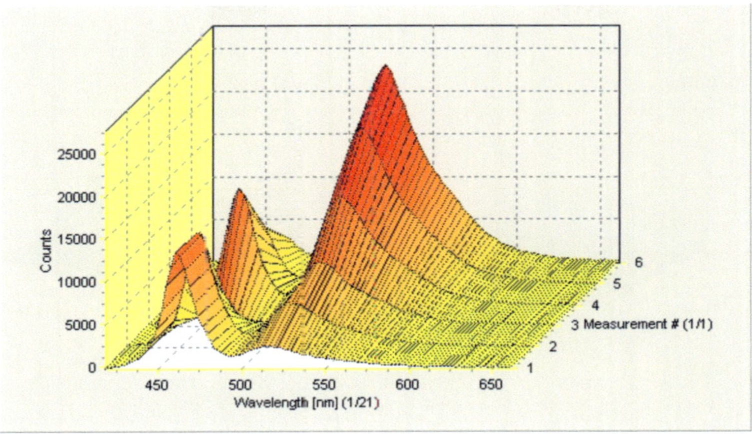

Figure 5-26: Fluorescence spectra of FITC-dextran detected by sensor 5 for the measurements at time points 0, 20, 100, 180, 255, 365 min

Illustration of the *in-situ* behavior of the molecules was prepared by the program MATLAB. Sensor numbers in the figure demonstrates the location of the sensors starting from cathode side. Therefore, the distribution of the cake over time can be evaluated from the fluorescence emission peaks (at 520 nm wave length) properly.

Emission spectra of the diagrams reveal the continuous increase in the concentration of the filter cake (see Figure 5-27). At first time point (t=0), the uniform distribution of concentration was viewed at all 5 sections resulting from the initial conditions. Similar uniform character of the distribution was visualized at 20 min with a slightly increased concentration. However, there was no labeled dextran more on the cathode side demonstrated by the decrease of the intensity of the first sensor. With passing time, concentration in the filter chamber increased continuously. At 100 min, still no high concentration gradient could be monitored.

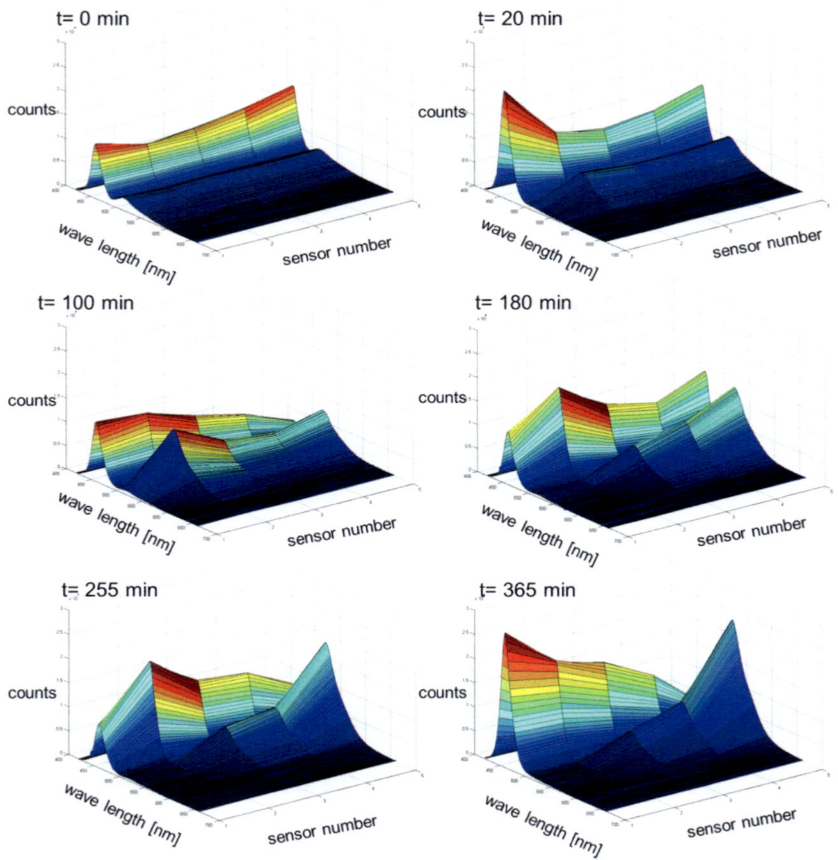

Figure 5-27: 3-D visualization of concentration gradient in the filter chamber at various time points (experimental conditions, 1 bar-80 V)

Beginning from 180 min, spectral analysis revealed a remarkable concentration gradient. As demonstrated on the last diagram in Figure 5-27 (at 365 min), the emission spectra represent a further impressive concentration gradient with more increased values in comparison to the ones characterized at 255 min.

The results demonstrate the gradual increase in concentration starting from cathode through anode side. Analysis of the *in-situ* behavior of the electrofiltration characteristics enables further improvement of electrofiltration in order to access the technology in industry. On the other hand, the reflection intensities fluctuated frequently and were affected strongly by environmental parameters. This was mainly based on the increasing concentration gradient. The sensors had two channels which were 200 µm close to each other. One of them picked the light, the other one emitted. When the concentration was low, picking and emitting function of the sensor worked well. However, when the concentration increased, the light could not go far away in the dispersion due to the short way of the light. As a result, most of the light could not reach the second channel. Therefore, reflection peak intensities remained at low values. Until the fourth sensor, concentration was not too high and reflection intensity could be detected properly. Nevertheless, in the fifth section where filter cake accumulated, the described phenomenon occurred and reflection peaks revealed low intensities. In addition to the explained phenomenon the reflection peaks were influenced by other side effects such as mixing effects of labeled and unlabeled biopolymer, different molecular weight of the substances and sliding down of the filter cake on membranes.

6 Future aspects

Future aspects of electrofiltration can be envisaged by facing the drawbacks and analyzing the current experimental results. This chapter will give an overview of further approaches which make technical improvements for the industria implementation of electrofiltration.

6.1 *In-situ* desalination

As previously discussed in chapters 4 and 5, the high salinity cf biopolymer dispersions increases the energy requirements and decreases the filtration rates compared with the dispersion with lower salinity. Therefore, integration of desalination system into the electrofiltration offers a valuable contribution.

Pre-experiments were performed aiming to analyze the feasibility of the *in-situ* desalination of filtrate solution. Experimental set up was changec with additional vessels and pumps for the exchange of filtrate solution Figure 6-1 presents the integrated desalination system with conventiona electrofiltration implementation. The biopolymer dispersion was transported continuously into the filter chamber under overpressure. Electric field was applied by generator and the filtrate solution was collected in the filtrate vessel together with the buffer solution. Online data acquisition was provided by the computer system and in periodical time intervals the filtrate solution in the vessel was exchanged with distilled water. The amounts of the pumped filtrate solution and the distilled water were kept equal in order to obtain the same dilution ratio. As a result of the exchange strategy, the conductivity of the filtrate solution decreased. Experiments were conducted by using xanthan dispersion and phosphate buffer with equal conductivity.

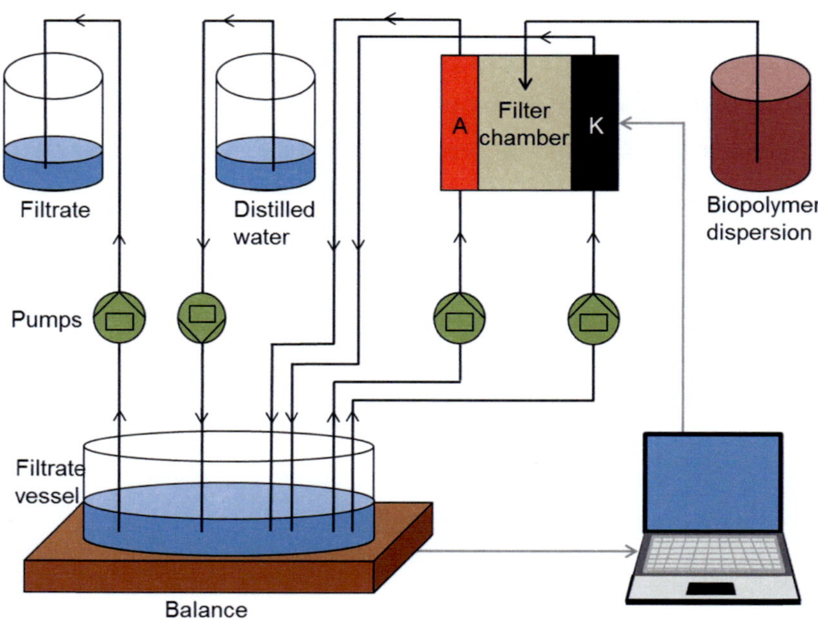

Figure 6-1: *In-situ* desalination with electrofiltration system

Every 30 minutes filtrate solution was pumped out during the experiments. 1 bar overpressure was applied with various applied voltages. Experiment results revealed that lowering the conductivity in filtrate vessel reduced the passing current through the filter chamber. Current values decreased by the exchange of the filtrate depending on the intensity of the applied voltages. Without exchanging the filtrate, the average current values were 12 mA and 22 mA for 40 V and 80 V applied voltages, respectively. With exchanging the filtrate, the average current values were reduced to 9.5 mA and 17 mA (see Figure 6-2).

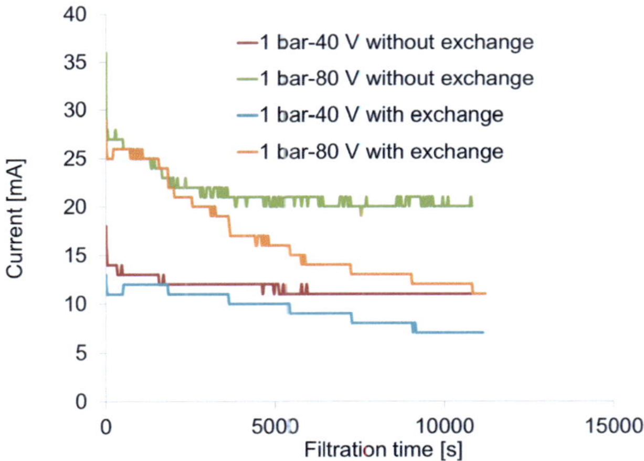

Figure 6-2: The current drop for the experiments with and without exchange of the filtrate

On the other hand, the decrease in passing current does not directly reduce the production costs. When energy consumption grades were calculated per filtrate mass, it was observed that the required energy to filtrate 100 g of filtrate increased with the exchange of filtrate during the experiments (see Figure 6-3). Moreover, the difference in energy consumption grades between with and without exchange of the filtrate solution increased when higher voltage was applied. The results were caused by the filtration kinetics of the experiments. Depending on the conductivity of the filtrate which was reduced step by step with the exchange of the filtrate, lower filtrate mass was achieved compared with the experiments without filtrate exchanging process (see appendix A4).

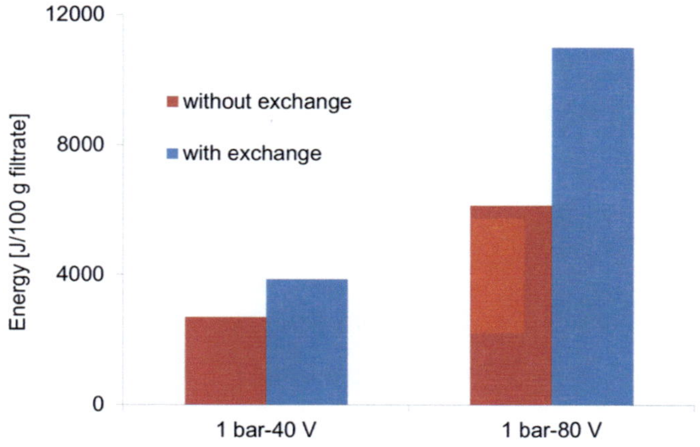

Figure 6-3: Required energy in joules per 100 g filtrate for the experiments with and without exchange of the filtrate

These results correlate well with the results discussed in sections 5.1 and 5.2. The slightly higher conductivity of the buffer solution compared with the conductivity of the biopolymer dispersion in the filter chamber leads to an improvement in filtration kinetics until a product related specific conductivity of buffer solution was applied (see section 5.2). However, as discussed in section 5.1, the use of lower conductivity in the filtrate vessel compared with the biopolymer dispersion conductivity in the filter chamber resulted in a decrease in the gradient of filtrate mass due to the highly change of ionic interactions on the membrane surfaces.

Thermodynamic instability entails the interfacial electrochemical phenomena at membranes and affects the transfer process. Consequently, the required energy to filtrate the same amount of biopolymer dispersion increases. Since the salinity of the filtrate solution is a critical factor, as a future approach of desalination technology, the time intervals for the exchanging of the buffer solution can be determined depending on the online measured conductivity of the filtrate solution. Controlled exchange strategy could prevent the reduction of the filtrate mass gradient allowing high efficiency with economic feasibility.

6.2 Optimization of the filter chamber geometry

6.2.1 Electrocandle filter

Electrocandle filter is an electrofiltration chamber system which has a cylindrical form (see illustration in Figure 6-4). The system provides decreased energy consumption (up to 40%) in comparison to the plate electrofiltration system (pilot-scale set-up) due to the radial form related energy calculations (Hofmann 2005). The electrocandle system works with same logic as the other electrofiltration systems. The biopolymer solution is filled between anode (outer) and cathode (inner) polyethersulfone concentric membranes which have different diameters. Cathode membrane works as working membrane and the room between electrodes and membranes are flushed by using buffer solution.

Filtration kinetics of xanthan dispersion with an initial concentration of 5 g/L was presented in Figure 6-5 in the presence and absence of electric field. The filtrate mass increased almost 4 times by application of electric field compared with the experiment result without application of electric field. Despite the increase of filtrate flux, some circumstances have to be faced.

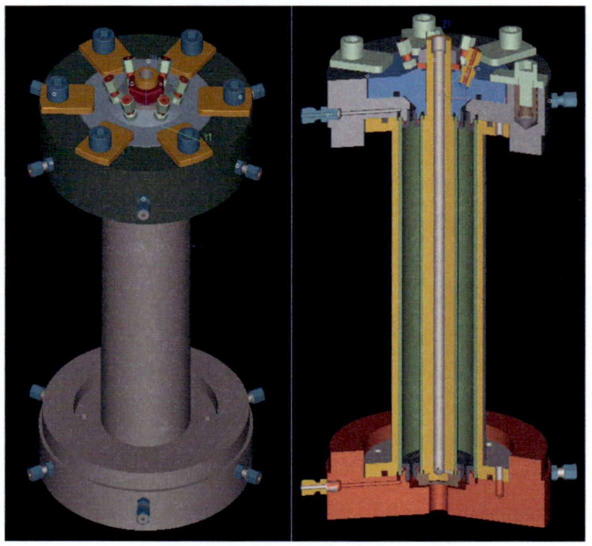

Figure 6-4: Electrocandle filter external (left) and cross sectional (right) appearance

In order to have a cylindrical membrane form that fits into the chamber, the reciprocal edges of the flat membranes were compounded together by using an adhesive. However, the leakage at connection places hinders the efficiency of experiments. These critical places are the linear overlapping adhesion zone of the membrane edges and the top/bottom of the electrocandle where membranes are attached to O-rings. Several adhesives were tested in order to find the most appropriate one and some adhesives revealed good results. However, obtaining flatness at the overlapping surface of the membrane is a difficult task due to the deforming effect of adhesives. The wavy membrane structure at the linear adhesion zone distorted the inhomogeneity of electric field and consequently energy consumption was higher compared with the theoretically assumed values.

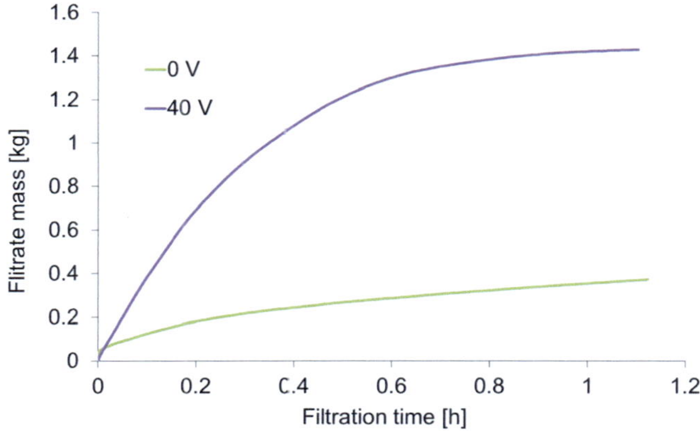

Figure 6-5: The gradient of filtrate mass of xanthan dispersion in the presence and absence of applied electric field when using electrocandle filter

A new method to compound the reciprocal membrane edges has to be developed (e.g welding) or as a promising alternative to the flat membranes, ceramic membranes can offer good properties such as original cylinder form, rigidity of the material and respectively an easier constructing.

6.2.2   Low voltage electrofiltration

Occurrence of water electrolysis is one of the most important drawbacks of the conventional electrofiltration system. As electrolysis occurs, hydrogen is produced at the cathode (see reaction 6.1) and oxygen is produced at the anode (see reaction 6.2). The amount of generated hydrogen and oxygen is proportional to the total electrical charge conducted by the biopolymer dispersion.

In order to understand the effect of electrolysis thoroughly, anode and cathode side flushing solutions (phosphate buffer) with filtrate solution were collected in separate vessels. The obtained data revealed that pH decreased from 7 to 2 in the anode side vessel due to the presence of H_3O^+ ions and increased from 7 to 10 in the cathode side vessel due to the presence of OH^- ions (see appendices A5 and A6).

<u>Anode reaction</u>:

$$6H_2O \rightarrow O_2(g) + 4H_3O^+ + 4e^- \qquad E_0 = 1.23\ V \qquad 6.1$$

<u>Cathode reaction</u>:

$$2H_2O + 2e^- \rightarrow H_2(g) + 2OH^- \qquad E_0 = -0.83\ V \qquad 6.2$$

Mass balance analysis demonstrated that ~30 g of water disappeared due to the electrolysis of water when the concentration of xanthan dispersion 28 times increased by application of 80 V. Therefore, a considerable part of applied electricity is used for the electrolysis. Moreover, the energy requirement increases when working with the high saline biopolymer dispersions.

The construction of a new designed electrofiltration chamber which has close distances between anode and cathode electrodes can solve the electrolysis problem. Application of adequate close distances hinders the occurrence of electrolysis, since the applied voltage will be lower compared with the standard electrode potential to electrolyze water. An illustration of low voltage electrofiltration system is presented in Figure 6-6 in terms of current-voltage diagram.

On the other hand, the selection of electrode material plays an important role for the system improvement. During this work, IrO_2 coated Ti anode electrodes and stainless steel cathode electrodes were used due to the relative low production grade of electrolysis products. Replacing metal oxide electrodes with new electrode materials can offer additional reduction of costs and therefore have to be further studied.

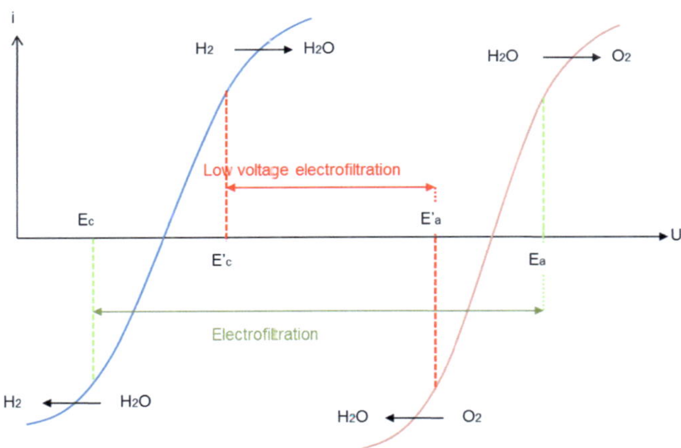

Figure 6-6: Current-Voltage diagram of low voltage electrofiltration

6.2.3 Continuous removal of filter cake

Continuous electrofiltration system involves the advent technology of filter cake removal. Formerly, filter cakes were discharged after performing experiments manually by opening the chambers. In order to install a continuous system, pre-experiments had to be first performed with the development of a semi-continuous process. Therefore, a filter chamber providing the removal of filter cake was designed and tested After the removal of the cake, the next filtration cycle commenced Several removal methods were tested including injection of compressed air (0.5 bar) into the top of the filter chamber, application of vacuum on the bottom of the filter chamber (16 mbar) and the combination of both methods simultaneously. Using vacuum as a removal method resulted in the damage of membranes and the partial removal of the cake. Analysis of the results demonstrated that the application of compressed air injection is more advantageous and has high optimization potential. Based on this, air injection system was improved. In addition to the central inlet which has 3 mm diameter for a dominant flow, two more side inlets (60° from the central one) were placed with a diameter of 1 mm. With these air inlets, removal of filter cake was achieved completely.

Investigations were revealed that overall concentration profile, viscosity and operation conditions influence the cake removal. Since the experiments were performed using the model product xanthan, the removal system has to be tested with other biopolymers in order to confirm the validity of the system. Moreover, further generation of electrofiltration has to be improved with the improvement of semi-continuous technology by application of continuous pressure difference ($\Delta P = 0.5$ bar) over the vertical access and outflow of the filter chamber.

6.3   Simulation

Mathematical model and numerical simulation method of electrofiltration was developed by using the program MATLAB.

Convection, diffusion, electrophoretic and reaction phenomena were taken into consideration for all the chemical compounds in the system including the biopolymer, dissociated biopolymer and the ions. Convection forces are related to the flow rate and differently than the previous studies flow rates were determined by Yukawa's equation (see equation 2.22) which defines the flow rate in consideration with the effect of applied electric field. Moreover, the filtration system was divided into 106 sub-cells for a better discretization. The chemicals were assumed to be high diluted and the transport velocities correspond to the ones of carrier fluid (no Stokes coefficient). The diffusion phenomenon was ruled by the first Fick's law and the diffusion coefficient was calculated by Einstein's equation for charged species (Hofmann 2005). The third phenomenon was electrophoretic forces and Manning's theory was considered (Manning 1981) for the calculations of mobility of the polymer with the considered ions Na^+ and Cl^-. The last phenomenon which leads to the variation of concentration was chemical reaction. In this work, polymer dissociation and the water dissociation were considered. Based on the model, simulation was performed and promising results were achieved. Figure 6-7 presents the comparison between the experimental and simulated filtrate mass-time curves by application of 1 bar and 40 V.

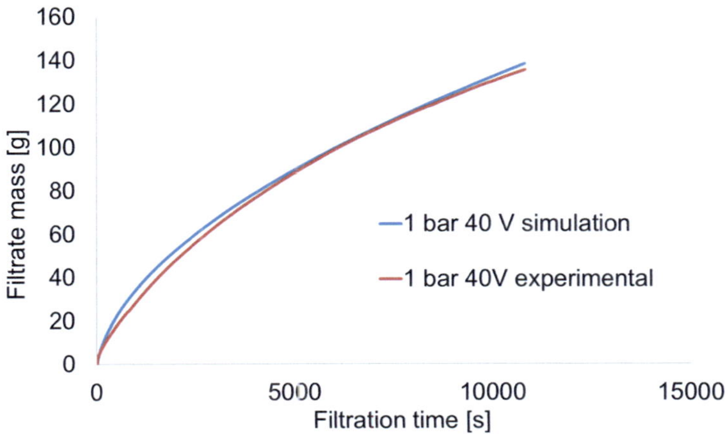

Figure 6-7: Comparison between simulated and experimented results at 40 V

As demonstrated, simulation result is very close to the experimental result. In addition, the simulation of the filter cake accumulation is demonstrated in Figure 6-8 in terms of concentration, time and position in the filter chamber. The left side of the figure corresponds to the cathode side and right side corresponds to the anode side filter cake accumulation.

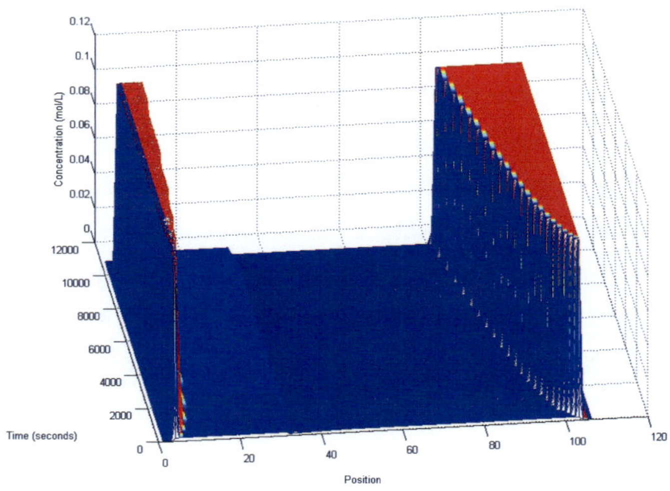

Figure 6-8: Filter cake composition in the filter chamber

However, by the increase of voltage, simulated data deviated slightly from experimented value (see appendix A7) since Manning's theory is valid only for high diluted biopolymer dispersions. Further approaches to identify the polymer behavior in viscous environments could help to improve the simulation system.

7 Summary and outlook

The rapid growth in biotechnology industry offers a remarkable interest in the production of biopolymers. Due to the challenges in the isolation and purification of biopolymers, downstream processing is a critical step which increases the production costs. Membrane processes are one of the most common separation processes in biotechnology that accomplish the process requirements such as high efficiency and selectivity. However, dead-end filtration of biopolymers is difficult since the biopolymer filter cakes have a highly compressible character. On the other hand, cross-flow filtration as an alternative technique can result in the disruption of chemical structure of sensitive biopolymers caused by the shear forces arising from tangential flow. In order to avoid limitations and decrease the downstream processing steps, integrative methods have to be developed considering the physicochemical properties of the biopolymers.

Electrofiltration is a separation method which combines dead-end filtration with electrophoresis for colloidal substances as biopolymers.

The application of electrofiltration in the separation and purification of biopolymers delivers promising results and offers a valuable technology for the production of biopolymers. The introduced electric field induces an electrophoretic flux of charged biopolymers. Filter cake formation occurs on the membrane at the side of the electrode which is oppositely charged to the product. The lack of thick filter cake on the other side results in a significant reduction of filtration time compared with conventional filtration technique. Within the frame of this work, the process availability for three of technically important biopolymers and the improvement of the electrofiltration strategy was investigated. Regarding the material based part of the investigations, filtration kinetics of PHB, chitosan and hyaluronic acid by variation of the process parameters were studied and the characterization of the products were done. For the improvement part of the research, dynamics of filter chamber system was analysed by several techniques.

PHB is a biodegradable plastic with the variety of special applications in medicine, agriculture and packaging industry. As an alternative to petroleum products, PHB presents ecological advantages. However, high costs originating from substrates and downstream processing costs hinder the commercial applications of PHB. Purification of PHB by means of electrofiltration presented an innovative approach in the production chain of the biopolymer. Electrofiltration was performed for the PHB containing dispersions obtained from batch and fed-batch processes after cell disruption. Up to 4 times increase of the concentration factor compared with the conventional filtration presented the definitive positive effect of applied electric field on the filtration kinetics of PHB dispersions. The high negative charge of the PHB molecules caused the improvement in kinetics. Different cultivation procedures and molecular weights of the products affected the concentration factors slightly. Moreover, using electrofiltration as an alternative technology demonstrated advantages concerning its high performance and cost efficiency. The energy reduction in comparison to the common method for the electrofiltration of PHB was in the range of 70-80% depending on the synthesis conditions. In addition, the energy required for electrofiltration experiments which were performed using the fed-batch produced dispersion was reduced by 17% compared with the experiment with constant conditions. Characterization analyses revealed the lack of unwanted modifications of PHB molecules. Therefore, the results demonstrated that the application of electrofiltration opens new possibilities for PHB production meeting the current downstream processing trends.

Chitosan is a biocompatible and biodegradable polysaccharide with applications in various fields, especially in cosmetics, pharmaceuticals and medicine. Cell walls of fungi or exoskeletons of arthropods are the sources of chitosan. However inevitably, the conventional separation methods have several drawbacks such as high generated costs and the possible structural changes during purification, which decrease the final quality. Fungi originated chitosan demonstrates heavy metal free character with low molecular weight which is attractive for medical applications and furthermore the production is not limited depending on the season. Therefore, it is important to improve the production process of fungi derived chitosan. Chitosan is a positively charged biopolymer with a high zeta potential enabling its purification by electrofiltration. In this work, electrofiltration was conducted to chitosan dispersions derived from crabs and fungi (*M. rouxii* and *A. coerulea*). The results revealed the advantages of application of electrofiltration in the chitosan production. The introduction of electrofiltration is attractive especially for fungi derived chitosan due to the high compressibility of the filter cakes. In comparison to conventional filtration, concentration factors based on the permeate mass of samples increased up to 3, 5 and 12 times for crabs, *M. rouxii* and *A. coerulea* derived chitosan dispersions respectively in the same period of time. The differences in concentration factors were based on the different dispersion compositions depending on the cultivation media. The limitations caused by the compressible character of the product were overcome by application of electric field. For the same amount of filtrate, electrofiltration was up to 15 times faster than conventional filtration for chitosan derived from *A. coerulea*. Moreover, electrofiltration required 91% less energy for the separation of crab-derived chitosan and 64% less energy for the separation of fungus-derived chitosan compared with precipitation method. The difference in energy requirements were based on the different conductivities, filtrate mass gradient and respective energy consumptions. Furthermore, the lack of structural change and the reduced heavy metal and protein content by application of electric field demonstrated the significant advantages of the technique which presents a valuable contribution to medical industry.

Hyaluronic acid is a biodegradable and biocompatible polysaccharide which has wide variety of medical and cosmetic applications. Hyaluronic acid is found in the form of a hydrated gel in human and animal tissues and additionally it can be synthesized by some bacteria. Rooster combs are the classical sources for the product however this source can cause inflammatory reactions after injection. Furthermore, ethic objectives are discussed. Purification of hyaluronic acid is mainly performed by extraction with organic solvents and precipitation with alcohol.

Nevertheless, this method not only increases the processing time and costs, but also leads to structural changes caused by organic solvents. The negative electric charge of hyaluronic acid enables the electrophoretic separation in a faster and more effective process with mild operational conditions. In comparison to conventional filtration, the concentration factor based on the permeate mass of samples was increased up to 4 times for the same experimental time for hyaluronic acid dispersions. Moreover, the application of electrofiltration presented no negative effect on the structure of HA molecules. These results demonstrated that electrofiltration is a relevant and promising technique in downstream processing of hyaluronic acid delivering pure product with desired unchanged specifics. The decrease in energy consumptions with increased filtration kinetics of all investigated products can be further improved by the optimization of electrofiltration system.

The second part of the work was stated for the analysis of the technical process and optimization of the system. The effect of conductivity in the filter chamber and flushing chambers was investigated. Ion chromatography analysis of filtrate samples was performed revealing the concentration and the variety of migrating ions. The experiments demonstrated that the conductivity of the buffer solution has a significant role on filtration kinetics. The difference in conductivity of biopolymer dispersion and buffer solution is related to the behavior of filtrate mass gradient. The most efficient results were received when the conductivity difference in and out of the filter chamber was not too high. Using a specific relative higher buffer conductivity compared with the biopolymer dispersion improved the filtration kinetics without increase of energy consumption for a specified filtrate mass. However, choosing the conductivity of buffer solution has to be decided by considering filtration behavior which is product related characteristic. For further process and device optimization, *in-situ* behavior of electrofiltration was investigated by two new designed filter chamber systems. The first system was constructed with integrative voltage sensors distributed in the filter and flushing chambers in order to measure and characterize the voltage drop through the system. The voltage drop on the filter media and the distribution of the applied voltage between the electrodes demonstrated promising information for the development of the system. Voltage drop near membranes was almost 34% more than the theoretical assumptions due to the filter cake accumulation and the resistance of filter media. This resulted in the unequal distribution of the voltage and as a result effective voltage remained 33% less compared with the theoretical calculations. The voltage drop on anode side increased with the increase of filter cake accumulation. In addition, using of filter media reduced the interior voltage by 8%. Different forms and/or materials of filter media can increase the effective voltage in the filter chamber.

The second system was designed with integrative fluorescence sensors distributed only in the filter chamber. The experiments were performed in order to evaluate the concentration gradient in the filter chamber by using the new designed chamber with fluorescence sensors. A LED-induced fluorescence device enabled monitoring of the migration of the Fluorescein isothiocyanate labeled dextran molecules providing 3-D visualization of the concentration gradient. The results demonstrated the concentration gradient from cathode side towards anode side membrane. Correlation between intensities and concentration revealed that on anode side concentration reached up to 10-fold compared with initial values for the tested time.

For the future aspects of electrofiltration several approaches were tested with pre-experiments. The experiments revealed that although the *in-situ* desalination of filtrate solution reduced the passing current, the required energy to filtrate the same amount of biopolymer dispersion increased. Therefore, for further development of *in-situ* desalination system, the time intervals of buffer solution-exchanging should be kept under control by online measuring of the conductivity in the filtrate vessel. On the other hand, some experiments were performed with model products by using an electrocandle system which has a reduced energy potential. The limitations arising from the sealing of the membrane were characterized and several adhesives were tested. Further investigations can be performed in order to seal the reciprocal membrane edges or other membrane materials such as ceramic membranes can be used due to the advantageous properties and an easier construction. One other important strategy to reduce the energy consumption was to reduce the distance between anode and cathode so that no electrolysis occurs. Pre-characterizations revealed the possibility of the application of the strategy. However, experiments have to be performed with a new chamber system. Last but not least, simulation of the electrofiltration was configured giving improved results. Simulation of the system can be further investigated for high concentrations in the filter chamber.

Zusammenfassung und Ausblick

Das schnelle Wachstum in der Biotechnologie-Industrie bietet ein beachtenswertes Potential bei der Produktion von Biopolymeren. Aufgrund der Herausforderungen bei der Isolierung und Aufreinigung von Biopolymeren, ist das Downstream-Processing ein entscheidender Schritt, der die Produktionskosten erhöht. Membranverfahren gehören zu den häufigsten Trennverfahren in der Biotechnologie, was die Prozessanforderungen wie hohe Effizienz und Selektivität betrifft. Allerdings ist die Dead-End Filtration von Biopolymeren schwierig, da der Biopolymer-Filterkuchen einen hoch kompressiblen Charakter hat. Andererseits kann die Cross-Flow-Filtration als alternative Technik die Zerstörung der chemischen Struktur der empfindlichen Biopolymere durch die Scherkräfte aus der Tangentialströmung verursachen. Um Einschränkungen zu vermeiden und die Aufarbeitungsschritte zu reduzieren, sollen integrative Methoden unter Berücksichtigung der physikalisch-chemischen Eigenschaften der Biopolymere entwickelt werden.

Elektrofiltration ist für kolloidale Substanzen wie Biopolymere ein Trennverfahren, das die Dead-End-Filtration mit Elektrophorese kombiniert. Die Anwendung der Elektrofiltration in der Separation und Aufreinigung von Biopolymeren liefert vielversprechende Ergebnisse und bietet eine wertvolle Technik für die Produktion von Biopolymeren. Das angelegte elektrische Feld induziert einen elektrophoretischen Fluss auf geladene Biopolymere. Der Filterkuchen wird auf der Membran an der Seite der Elektrode, die zu dem Produkt entgegengesetzt geladen ist, gebildet. Das Fehlen eines ausgeprägten Filterkuchens auf der anderen Seite führt zu einer deutlichen Reduzierung der Filtrationszeit im Vergleich mit herkömmlicher Filtrationstechnik. Im Rahmen dieser Arbeit wurden die Prozess-Verfügbarkeit für drei der technisch wichtigsten Biopolymeren und denen Prozess-Verbesserung mittels der Elektrofiltration untersucht. Bezüglich des stofflich orientierten Teils der Untersuchungen wurden die Filtrationskinetiken von PHB, Chitosan und Hyaluronsäure durch Variation der Prozessparameter untersucht und die Charakterisierung der Produkte durchgeführt. Für die Prozess-Analyse als Teil der Arbeit wurden die Dynamik des Filtrations-Systems und die räumliche Struktur durch verschiedene Methoden analysiert.

PHB ist ein biologisch abbaubarer Kunststoff mit einer Vielzahl von speziellen Anwendungen in Medizin, Landwirtschaft und Verpackungs-industrie. Als Alternative zu Erdölprodukten bietet PHB ökologische Vorteile. Allerdings behindern die hohen Kosten, die sich aus Substrat- und Aufarbeitungskosten zusammensetzen, die kommerziellen Anwendungen von PHB. Die Aufreinigung von PHB mittels Elektrofiltration stellt einen innovativen Ansatz in der Produktion des Biopolymers dar. Die Elektrofiltration wurde für die PHB haltige Dispersionen aus Batch-und Fed-Batch-Prozesse nach dem Zellaufschluss durchgeführt. Die bis zu vier-fache Erhöhung des Konzentrationsfaktors im Vergleich zu der konventionellen Filtration zeigte die definitiv positive Wirkung des angelegten elektrischen Feldes auf die Filtrationskinetik der PHB-Dispersionen. Die hohe negative Ladung der PHB-Moleküle verursachte die Verbesserung der Kinetik. Die Unterschiede in der Kultivierungsstrategie und bei den Molekular-gewichten haben die Konzentrationsfaktoren etwas beeinflusst. Die Elektrofiltration als eine alternative Technologie demonstrierte so Vorteile hinsichtlich ihrer hohen Leistungsfähigkeit und Kosteneffizienz

Die Senkung des Energieverbrauchs im Vergleich zu den üblichen Verfahren für die Elektrofiltration von PHB war im Bereich von 70-80%, je nach Synthesebedingungen. Außerdem wurde der Energieverbrauch für die Prozessführungs-Experimente, die unter Verwendung der mittels Fed-Batch produzierten Dispersion erforderlich waren, 17% im Vergleich zu konstanten Bedingungen reduziert. Bei Charakterisierungs-Analysen wurden keine unerwünschten Änderungen der PHB-Moleküle nachgewiesen. Daher zeigten die Ergebnisse, dass die Anwendung der Elektrofiltration neue Möglichkeiten für die PHB-Produktion eröffnet.

Chitosan ist ein biokompatibles und biologisch abbaubares Polysaccharid mit Anwendungen in verschiedenen Bereichen, insbesondere in der Kosmetik, Pharmazie und Medizin. Die Zellwände von Pilzen oder die Exoskelette von Arthropoden sind die möglichen Quellen von Chitosan. Allerdings haben die herkömmlichen Separationsmethoden einige Nachteile wie hohe Kosten und die möglichen strukturellen Veränderungen während der Aufreinigung, die die endgültige Qualität verringert. Pilze-Chitosan zeigt einen schwermetallfreien Charakter mit niedrigem Molekulargewicht, was attraktiv für medizinische Anwendungen ist und darüber hinaus nicht die Produktion abhängig von der Saison begrenzt. Daher ist es wichtig, den Produktionsprozess des von Pilzen gewonnenen Chitosans zu verbessern. Chitosan ist ein positiv geladenes Biopolymer mit hohem Zeta-Potential, was seine Separation durch Elektrofiltration ermöglicht. In dieser Arbeit, wurde die Elektrofiltration von Chitosan-Dispersionen aus Krabben und Pilzen (*M. rouxii* und *A. coerulea*) durchgeführt. Die Ergebnisse zeigten die Vorteile der Anwendung von Elektrofiltration bei der Chitosan-Produktion. Die Einführung der Elektrofiltration ist besonders für das aus Pilzen gewonnene Chitosan aufgrund der hohen Kompressibilität der

Filterkuchen attraktiv. Im Vergleich zur herkömmlichen Filtration wurden die auf die Permeatmasse bezogenen Konzentrationsfaktoren bis zu 3, 5 und 12 Mal für Krebse, *M. rouxii* und *A. coerulea* gewonnene Chitosan-Dispersionen erhöht. Die Unterschiede in den Konzentrations-faktoren resultieren aus den verschiedenen Dispersionszusammen-setzungen je nach Kultivierungsmedien. Die Einschränkungen, die durch den kompressiblen Charakter des Produkts verursacht waren, wurden durch Anwendung eines elektrischen Feldes beseitigt. Für die gleiche Filtratmasse war die Elektrofiltration von Chitosan aus *A. coerulea* bis zu 15-mal schneller als die herkömmliche Filtration. Außerdem benötigte die Elektrofiltration 91% weniger Energie für die Separation von aus Krabben gewonnenem Chitosan und 64% weniger Energie für die Separation von aus Pilzen gewonnenem Chitosan im Vergleich mit der Präzipitation. Der Unterschied in den Energie-Anforderungen wurde durch die verschiedenen Leitfähigkeiten und Filtratmassen-Gradienten begründet. Weiterhin stellt das Fehlen von strukturellen Veränderungen und die reduzierten Schwermetall- und Eiweißgehalte durch die Anwendung des elektrischen Feldes die Methode als einen wertvollen Beitrag zur medizinischen Industrie dar.

Hyaluronsäure ist ein biologisch abbaubares und biokompatibles Polysaccharid, das vielfältige medizinische und kosmetische Anwendungen hat. Hyaluronsäure ist in der Form eines hydratisierten Gels in den menschlichen und tierischen Geweben vorhanden und kann zusätzlich von einigen Bakterien synthetisiert werden. Hahnenkämme sind die klassischen Quellen für das Produkt jedoch kann die Verwendung dieser Quelle entzündliche Reaktionen nach der Injektion verursachen. Außerdem stehen ethische Bedenken in der Diskussion. Die Aufreinigung von Hyaluronsäure wird vor allem durch Extraktion mit organischen Lösungsmitteln und Fällung mit Alkohol durchgeführt. Dennoch erhöht diese Methode nicht nur die Bearbeitungszeit und die Kosten, sondern führt auch zu strukturellen Veränderungen durch organische Lösungsmittel. Die negative elektrische Ladung von Hyaluronsäure ermöglicht die elektrophoretische Separation als ein schnelleres und effektiveres Verfahren mit milden Betriebsbedingungen. Im Vergleich zur herkömmlichen Filtration wurde der auf die Permeatmasse bezogene Konzentrationsfaktor für Hyaluronsäure-Dispersionen bis zu 4-mal bei gleichen Versuchsbedingungen erhöht. Die Anwendung von Elektrofiltration bewirkt keine negativen Auswirkungen auf die Struktur der Hyaluronsäure-Moleküle. Diese Ergebnisse zeigten, dass die Elektrofiltration eine relevante und vielversprechende Technik in der Weiterverarbeitung von Hyaluronsäure mit den gewünschten unveränderten Spezifikationen ist. Die Energieverbräuche aller untersuchten Produkte können durch die Optimierung des Elektrofiltrations-System noch verbessert werden.

Der zweite Teil der Arbeit wurde der Analyse des technischen Prozesses und der Optimierung des Systems gewidmet. Der Einfluss der Leitfähigkeit in der Filterkammer und Spülkammern wurde untersucht. Ionenchromatographische Analysen der Filtratproben wurden durchgeführt, um die Konzentration und die Art der wandernden Ionen zu bestimmen. Die Experimente zeigten, dass die Leitfähigkeit der Puffer-Lösung eine wichtige Rolle für Filtrationskinetik spielt. Das Verhalten des Filtratmasse-Gradienten hängt von der Differenz in der Leitfähigkeit der Biopolymer-Dispersion und Pufferlösung ab. Die effizientesten Ergebnisse wurden erhalten, wenn die Leitfähigkeits-differenzen zwischen innen und außen der Filterkammer nicht zu hoch sind. Mit Hilfe einer relativ höheren Puffer Leitfähigkeit im Vergleich zur

Biopolymer Dispersion verbessert sich die Filtrationskinetik ohne Erhöhung des spezifischen Energieverbrauchs. Die Leitfähigkeit der Pufferlösung soll also durch Berücksichtigung des Filtrationsverhaltens bzw. durch die Produktcharakteristik gewählt werden. Für die weitere Prozessoptimierung wurde das *in-situ* Verhalten der Elektrofiltration durch zwei neu entwickelte Filter Kammer-Systeme untersucht. Das erste System wurde mit verteilten integrativen Spannungssensoren in der Filter- und den Spülkammern aufgebaut, um den Spannungsabfall im System zu charakterisieren. Der Spannungsabfall über die Filtermedien und die Verteilung der angelegten Spannung zwischen den Elektroden zeigten vielversprechende Informationen zur Entwicklung der Filtrationsdynamik. Spannungsabfälle in der Nähe von Membranen waren fast 34% höher als die theoretischen Annahmen aufgrund der Filterkuchenbildung und des Widerstands der Filtermedien. Dies führte zu einer ungleichen Verteilung der Spannung und als Folge zur Reduzierung der effektiven Spannung um 33% im Vergleich mit den theoretischen Berechnungen. Der Spannungsabfall an der Anodenseite stieg mit der Erhöhung der Filterkuchenbildung. Darüber hinaus reduzierte die Verwendung von Filtermedien die innere Spannung über 8%. Verschiedene Formen und/oder Materialien von Filtermedien können die effektive Spannung in der Filterkammer erhöhen.

Das zweite System wurde mit nur in der Filterkammer verteilten integrativen Fluoreszenzsensoren aufgebaut. Die Experimente wurden durchgeführt, um den Konzentrationsgradient in der Filterkammer mit Fluoreszenz-Sensoren zu bewerten. Ein mit LED-Anregung ausgestattetes Fluoreszenz-Gerät ermöglichte die Beobachtung der Migration der Fluoresceinisothiocyanat-markierten Dextran-Moleküle und führte zur 3-D-Visualisierung des Konzentrationsgradienten. Die

Ergebnisse zeigten, dass die Konzentrationsgradient steigte von der kathodenseitigen nach anodenseitiger Membran. Die Korrelation zwischen Intensität und Konzentration demonstrierte, dass die Konzentration auf Anodenseite bis zu 10-fachen im Vergleich zu ursprünglichen Wert für die getestete Zeit erreichte.

Weiterführende Aspekte der Elektrofiltration wurden in mehreren Ansätzen durch Vorexperimente getestet. Die Experimente zeigten, dass obwohl die *in-situ* Entsalzung der Filtratlösung den Stromverbrauch reduzierte, die insgesamt benötigte Energie pro Filtratmenge erhöht wurde. Daher sollen für die weitere Entwicklung der *in-situ* Entsalzungsanlage die zeitlichen Abstände des Pufferlösungs-Austauschs durch Regelung der Leitfähigkeit im Filtratbehälter optimiert werden. Auf der anderen Seite wurden einige Versuche mit dem Elektrokerzen-System, welches ein reduziertes Energie-Potential hat, durchgeführt. Die Einschränkungen, die sich durch die Abdichtung der Membran ergaben, wurden charakterisiert und mehrere Klebstoffe getestet. Weitere Untersuchungen können durchgeführt werden, um die gegenüber liegenden Membrankanten zu dichten oder andere Membranmaterialien wie keramische Membranen aufgrund der vorteilhaften Eigenschaften und einer einfacheren Konstruktion zu testen. Eine weitere wichtige Strategie, um den Energieverbrauch zu reduzieren, ist es den Abstand zwischen Anode und Kathode zu reduzieren, so dass keine Elektrolyse stattfindet. Vorläufige Charakterisierungen zeigten die Möglichkeit der Anwendung dieser Strategie. Versuche sollen jedoch mit einem neuen Kammer-System durchgeführt werden. Nicht zuletzt, wurde die Simulation der Elektrofiltration konfiguriert und gute Ergebnisse erzielt und dargestellt.

Simulation des Systems können weiter für hohe Konzentrationen in der Filterkammer untersucht werden.

References

Alkrad JA, Mrestani Y, Stroehl D, Wartewig S, Neubert R (2003) Characterization of enzymatically digested hyaluronic acid using NMR, Raman, IR, and UV-Vis spectroscopies. Journal of Pharmaceutical and Biomedical Analysis 31(3):545-550.

Anderson AJ, Dawes EA (1990) Occurrence, metabolism, metabolic role, and industrial uses of bacterial polyhydroxyalkanoates. Microbiological Reviews 54(4):450-472.

Andre P (2004) Hyaluronic acd and its use as a rejuvenation agent in cosmetic dermatology. Seminars in Cutaneous Medicine and Surgery 23(4):218-222.

Anlauf H (1994) Standardfiltertests zur Bestimmung des Kuchen- und Filtermediumwiderstandes bei der Feststoffabtrennung aus Suspensionen (Teil 1). Filtrieren und Separarieren 8:2.

Aranaz I, Mengibar M, Harris R, Panos I, Miralles B, Acosta N, Galed G, Heras A (2009) Functional characterization of chitin and chitosan. Current Chemical Biology 3(2):203-230.

Arkhangelsky E, Kuzmenko D, Gitis V (2007) Impact of chemical cleaning on properties and functioning of polyethersulfone membranes Journal of Membrane Science 305(1-2):176-184.

Balazs N, Sipos P (2007) Limitations of pH-potentiometric titration for the determination of the degree of deacetylation of chitosan. Carbohydrate Research 342(1):124-130.

Bargeman G, Houwing J, Recio I, Koops GH, van der Horst C (2002) Electro-membrane filtration for the selective isolation of bioactive peptides from an alpha(s2)-casein hydrolysate. Biotechnology and Bioengineering 80(6):599-609.

Barnes GT, Gentle IR (2005) Interfacial Science. Oxford University Press Inc., New York.

Bartnicki-Garcia S, Nickerson WJ (1962) Isolation, composition, and structure of cell walls of filamentous and yeast-like forms of *Mucor rouxii*. Biochimica Et Biophysica Acta 58(1):102-119.

Bell G, Cousins RB (1994) Membrane separation processes, in: Weatherley LR (Ed.), Engineering Processes for Bioseparations. Butterworth-Heinemann Ltd., Oxford, pp. 135-165.

Blank LM, McLaughlin RL, Nielsen LK (2005) Stable production of hyaluronic acid in Streptococcus zooppidemicus chemostats operated at high dilution rate. Biotechnology and Bioengineering 90(6):685-693.

Boas NF (1949) Isolation of hyaluronic acid from the cocks comb. Journal of Biological Chemistry 181(2):573-575.

Bowen WR, Ahmad AL (1997) Pulsed electrophoretic filter-cake release in dead-end membrane processes. AIChE Journal 43(4):959-970.

Bowen WR, Cao XW, Williams PM (1999) Use and elucidation of biochemical data in the prediction of the membrane separation of biocolloids. Proceedings of the Royal Society of London Series a-Mathematical Physical and Engineering Sciences 455(1988):2933-2955.

Bowen WR, Doneva TA, Stoton JAG (2002) Protein deposition during cross-flow membrane filtration: AFM studies and flux loss. Colloids and Surfaces B-Biointerfaces 27(2-3):103-113.

Bowen WR, Williams PM (1996) Dynamic ultrafiltration for proteins – A colloidal interaction approach. Biotechnology and Bioengineering 50:125-135.

Brisson G, Britten M, Pouliot Y (2007) Electrically-enhanced cross-flow microfiltration for separation of lactoferrin from whey protein mixtures. Journal of Membrane Science 291(1-2):206-216.

Brors A (1992) Untersuchungen zum Einfluss von elektrischen Feldern bei der Querstromfiltration von biologischen Suspensionen. Fortschrittberichte VDI 284.

Cai J, Yang JH, Du YM, Fan LH, Qiu YF, Li J, Kennedy JF (2006) Enzymatic preparation of chitosan from the waste *Aspergillus niger* mycelium of citric acid production plant. Carbohydrate Polymers 64(2):151-157.

Canizares P, Saez C, Sanchez-Carretero A, Rodrigo MA (2009) Synthesis of novel oxidants by electrochemical technology. Journal of Applied Electrochemistry 39(11):2143-2149.

Caplan SR, Miller IR, Milazzo G (1995) Biolectrochemistry: General Introduction. Birkhäuser Verlag, Basel.

Chatterjee S, Adhya M, Guha AK, Chatterjee BP (2005) Chitosan from *Mucor rouxii*: production and physico-chemical characterization. Process Biochemistry 40(1):395-400.

Chen GGQ (2005) Polyhydroxyalkanoates, in: Smith R (Ed.), Biodegradable polymers for industrial applications, CRC Press, Boca Raton, pp. 32-56.

Chen SJ, Chen JL, Huang WC, Chen HL (2009) Fermentation process development for hyaluronic acid production by Streptococcus zooepidemicus ATCC 39920. Korean Journal of Chemical Engineering 26(2):428-432.

Chien LJ, Lee CK (2007) Enhanced hyaluronic acid production in Bacillus subtilis by coexpressing bacterial hemoglobin. Biotechnology Progress 23(5):1017-1022.

Chong BF, Blank LM, Mclaughlin R, Nielsen LK (2005) Microbial hyaluronic acid production. Applied Microbiology and Biotechnology 66(4):341-351.

Chung YC, Su YP, Chen CC, Jia G, Wang HI, Wu JCG, Lin JG (2004) Relationship between antibacterial activity of chitosan and surface characteristics of cell wall. Acta Pharmacologica Sinica 25(7):932-936.

Clarinval AM, Halleux J (2005) Classification of biodegradable polymers, in: Smith R (Ed.), Biodegradable polymers for industrial applications, CRC Press, Boca Raton, pp. 3-31.

Darcy H (1856) Les Fontaines Publiques de la Ville de Kijon. Victor Dalmont, Paris.

Delgado ÁV, Shilov VN (2006) Electrokinetics of suspended solid colloid particles. in: Somasundan P (Ed.), Encyclopedia of Surface and Colloid Science, Vol.3. CRC Press, Boca Raton, pp. 2233-2261.

Derjaguin BV, Landau L (1941) Theory of the stability of strongly charged lyphobic sols and of the adhesion of strongly charged particles in solutions of electrolytes. Acta Physicochim URSS 14:633-662.

Doi Y, Steinbüchel A (2002) Polyesters, in Steinbüchel A (Ed.), Biopolymers Vol. 3a-3c, Wiley-VCH, Weinheim.

Dörfler HD (2002) Grenzflächen und kolloid-disperse Systeme. Springer-Verlag, Heidelberg.

Duan XJ, Yang L, Zhang X, Tan WS (2008) Effect of oxygen and shear stress on molecular weight of hyaluronic acid produced by Streptococcus zooepidemicus. Journal of Microbiology and Biotechnology 18(4):718-724.

Dumitriu S (1998) Polysaccharides – structural Diversity and Functional Versatility. Marcel Dekker Inc., New York.

Enevoldsen AD, Hansen EB, Jonsson G. (2007) Electro-ultrafiltration of industrial enzyme solutions. Journal of Membrane Science 299(1-2):28-37.

European Bioplastics (2007) European Bioplastics boom in 2006 – Excellent outlook but more investment required to expand capacity. Plastics Engineering 63(1):6-7.

Fiorese ML, Freitas F, Pais J, Ramos AM, de Aragao GMF, Reis MAM (2009) Recovery of polyhydroxybutyrate (PHB) from *Cupriavidus necator* biomass by solvent extraction with 1,2-propylene carbonate. Engineering in Life Sciences 9(6):454-461.

Gao XD, Katsumoto T, Onodera K (1995) Purification and characterization of chitin deacetylase from *Absidia coerulea*. Journal of Biochemistry 117(2):257-263.

García AA, Bonen MR, Ramírez-Vick J, Sadaka M, Vuppu A (1999) Bioseparation Process Science. Blackwell Science, Inc., Massachusetts.

Ghirisan A, Hofmann R, Posten C (2005) Druckfiltration und Druck-Elektrofiltration von Hefesuspensionen. Filtrieren und Separieren 19(3):118-122.

Ghosh R (2006) Principles of Bioseparations Engineering. World Scientific Publishing Co. Pte. Ltd., Singapore.

González-Caballero F, Shilov VN (2006) Electrical double layer at a colloid particle, in: Somasundan P (Ed.), Encyclopedia of Surface and Colloid Science, Vol.3, CRC Press, Boca Raton, pp. 1932-1936.

Gözke G, Kirschhöfer F, Heissler S, Trutnau M, Brenner-Weiss G, Ondruschka J, Obst U, Posten C (2011) Filtration kinetics of chitosan separation by electrofiltration. Biotechnology Journal 7(2):262-274.

Gözke G, Posten C (2010) Electrofiltration of biopolymers. Food Engineering Reviews 2(2):131-146.

Green DW, Perry RH (2008) Perry's chemical engineers' handbook. McGraw-Hill Companies, Inc., New York.

Groeger G, Geyer W, Bley T, Ondruschka J (2006) Fermentative Herstellung von Chitosan aus Pilzmycelien. Chemie Ingenieur Technik 78(4):479-483.

Hakimhashem M, Saveyn H, De Bock B, Van der Meeren P (2010) Dead-end liposomal electro-filtration: Phenol removal by dioctadecyl dimethyl ammonium chloride as a case study. Chemical Engineering & Technology 33(8):1321-1326.

Hamann CH, Hamnett A, Vielstich W (2007) Electrochemistry. Wiley-VCH, Weinheim.

Hänggi UJ (1990) Pilot scale production of PHB in *Alcaligenes latus*, in Dawes EA (Ed.), Novel biodegradable microbial polymers. Kluwer Academic Publishers, Dodrecht, pp. 65-70.

Harrison RG, Todd P, Rudge SR, Petrides DP (2003) Bioseparations Science and Engineering. Oxford University Press, Inc., New York.

Hazer B, Steinbüchel A (2007) Increased diversification of polyhydroxyalkanoates by modification reactions for industrial and medical applications. Applied Microbiology and Biotechnology 74(1):1-12.

Henry JD, Lawler LF, Kuo CHA (1977) Solid-liquid separation process based on cross flow and electrofiltration. Aiche Journal 23(6):851-859.

Hidalgo-Alvarez R, Martin A, Fernandez A, Bastos D, Martinez F, Nieves FJ (1996) Electrokinetic properties, colloidal stability and aggregation kinetics of polymer colloids. Advances in Colloid and Interface Science 67:1-118.

Hiemenz PC, Rajagopalan R (1997) Principles of Colloid and Surface Chemistry. Marcel Dekker, Inc., New York.

Hocking PJ, Marchessault RH (1998) Introduction to biopolymers from renewable resources, in: Kap an DL (Ed.), Biopolymers from Renewab e Resources, Springer Verlag, Berlin, pp. 220-248.

Hofmann R (2005) Prozesstechnische Entwicklung der Presselektrofiltration als innovatives Verfahren zur Abtrennung von Biopolymeren (Dissertation), VDI Verlag, Düsseldorf.

Hofmann R, Kappler T, Posten C (2006) Pilot-scale press electrofiltration of biopolymers 51(3):303-309.

Hofmann R, Posten C (2003) Improvement of dead-end filtration of biopolymers with pressure electrofiltration. Chemical Engineering Science 58(17):3847-3858.

Hofmann R, Weber K, Herrenbauer M, Posten C (2001) Press electrofiltration - A highly promising method of bioseparation. Chemie Ingenieur Technik 73(9):1218-1224.

Hunter RJ (1993) Introduction to modern colloid science. Oxford University Press Inc., Oxford.

Hutterer KM, Jorgenson JW (2005) Separation of hyaluronic acid by ultrahigh-voltage capillary gel electrophoresis. Electrophoresis 26(10):2027-2033.

Ignatova EU, Gurov AN (1990) Principles of extraction and purification of hyaluronic-acid. Khimiko-Farmatsevticheskii Zhurnal 24(3):42-46.

Imam SH, Gordon SH, Shogren RL, Tosteson TR, Govind NS, Greene RV (1999) Degradation of starch-poly(beta-hydroxybutyrate-co-beta-hydroxyvalerate) bioplastic in tropical coastal waters. Applied and Environmental Microbiology 65(2):431-437.

Iritani E (2003) Properties of filter cake in cake filtration and membrane filtration. Kona 21:1-39.

Iritani E, Mukai Y, Murase T (1995) Properties of filter cake in dead-end ultrafiltration of binary protein mixtures with retentive membranes. Chemical Engineering Research & Design 73(A5): 551-558.

Iritani E, Nagaoka H, Katagiri N (2008) Determination of filtration characteristics of yeast suspension based upon multistage reduction in cake surface area under step-up pressure conditions. Separation and Purification Technology 63(2):379-385.

Iritani E, Ohashi K, Murase T (1992) Analysis of filtration mechanism of dead-end electroultrafiltration for proteinaceous solutions. Journal of Chemical Engineering of Japan 25(4):383-388.

Iwata M, Igami H, Murase T, Yoshida H (1991) Combined operation of electroosmotic dewatering and mechanical expression. Journal of Chemical Engineering of Japan 24(3):399-401.

Jacobson KS, Drew DM, He Z (2011) Efficient salt removal in a continuously operated upflow microbial desalination cell with an air cathode. Bioresource Technology 102(1):376-380.

Jacquel N, Lo CW, Wei Y H, Wu HS, Wang SS (2008) Isolation and purification of bacterial poly(3-hydroxyalkanoates). Biochemical Engineering Journal 39(1):15-27.

Jendrossek D, Handrick R (2002) Microbial degradation of polyhydroxyalkanoates Annual Review of Microbiology 56:403-432.

Jiang XA, Chen LR, Zhong W (2003) A new linear potentiometric titration method for the determination of deacetylation degree of chitosan. Carbohydrate Polymers 54(4):457-463.

194

John ME, Keller G (1996) Metabolic pathway engineering in cotton: Biosynthesis of polyhydroxybutyrate in fiber cells. Proceedings of the National Academy of Sciences of the United States of America 93(23):12768-12773.

Johnson K, Kleerebezem R, van Loosdrecht MCM (2010) Influence of ammonium on the accumulation of polyhydroxybutyrate (PHB) in aerobic open mixed cultures. Journal of Biotechnology 147(2):73-79.

Jones SA, Goodall DM, Cutler AN, Norton IT (1987) Application of conductivity studies and polyelectrolyte theory to the conformation and order-disorder transition of xanthan polysaccharide. European Biophysics Journal with Biophysics Letters 15(3):185-191.

Kakehi K, Kinoshita M, Yasueda S (2003) Hyaluronic acid: separation and biological implications. Journal of Chromatography B-Analytical Technologies in the Biomedical and Life Sciences 797(1-2):347-355.

Kanani DA, Sun XH, Ghosh R (2008) Reversible and irreversible membrane fouling during in-line microfiltration of concentrated protein solutions. Journal of Membrane Science 315(1-2):1-10.

Kaplan DL (1998) Introduction to biopolymers from renewable resources, in: Kaplan DL (Ed.), Biopolymers from Renewable Resources, Springer Verlag, Berlin, pp. 1-29.

Käppler T, Posten C (2007) Fractionation of proteins with two-sided electro-ultrafiltration. Journal of Biotechnology 128(4):895-907.

Kawaguchi Y, Doi Y (1990) Structure of native poly(3-Hydroxybutyrate) granules characterized by X-Ray-diffraction. Fems Microbiology Letters 70(2):151-156.

Khor E, Lim LY (2003) Implantable applications of chitin and chitosan. Biomaterials 24(13):2339-2349.

Khosravi-Darani K, Vasheghani-Farahani E, Shojaosadati SA, Yamini Y (2004) Effect of process variables on supercritical fluid disruption of *Ralstonia eutropha* cells for poly(R-hydroxybutyrate) recovery. Biotechnology Progress 20(6):1757-1765.

Kim SJ, Park SY, Kim CW (2006) A novel approach to the production of hyaluronic acid by *Streptococcus zooepidemicus*. Journal of Microbiology and Biotechnology 16(12):1849-1855.

Kissa E (1999) Dispersions – Characterisation, Testing, and Measurement. Marcel Dekker Inc., New York.

Koehler JA, Ulbricht M, Belfort G (1997) Intermolecular forces between proteins and polymer films with relevance to filtration. Langmuir 13(15):4162-4171.

Koller M, Atlic A, Gonzalez-Garcia Y, Kutschera C, Braunegg G (2008) Polyhydroxyalkanoate (PHA) biosynthesis from whey lactose. Macromolecular Symposia 272:87-92.

Kornyshew AA, Spohr E, Vorotyntsev MA (2002) Elecrochemical interfaces: at the border line. in: Gileadi E and Urbakh M (Ed.), Thermodynamics and Electrified Interfaces, Vol 1, pp. 33-132.

Kumar MNVR (2000) A review of chitin and chitosan applications. Reactive & Functional Polymers 46(1):1-27.

Larue O, Vorobiev E (2004) Sedimentation and water electrolysis effects in electrofiltration of kaolin suspension. AIChE Journal 50(12):3120-3133.

Lewis MJ (1996) Pressure-activated membrane processes, in: Grandison AS and Lewis MJ (Ed.), Separation Processes in the Food and Biotechnology Industries. Woolhead Publishing Ltd., Cambridge, pp. 65-96.

Lockhart, NC (1992) Combined field dewatering – Bridging the science-industry. GapDrying Technology 10(4):839-874.

Madison LL, Huisman G W (1999) Metabolic engineering of poly(3-hydroxyalkanoates): From DNA to plastic. Microbiology and Molecular Biology Reviews 63(1):21-53.

Manegold E (1937) The effectiveness of filtration, dialysis, electrolysis and their intercombinations as purification processes. Transactions of the Faraday Society 33:1088-1094.

Manning GS (1981) Limiting laws and counterion condensation in poly-electrolyte solutions 7. Electrophoretic mobility and conductance. Journal of Physical Chemistry 85(11):1506-1515.

Medronho RA (2003) Solid-liquid separation, in: Hatti-Kaul R, Mattiasson B (Ed.), Isolation and Purification of Proteins, Marcel Dekker, Inc. New York, pp. 117-174.

Mehanna M, Saito T, Yan JL, Hickner M, Cao XX, Huang X, Logan BE (2010) Using microbial desalination cells to reduce water salinity prior to reverse osmosis. Energy & Environmental Science 3(8):1114-1120.

Meyer K, Palmer JW (1934) The polysaccharide of the vitreous humor. Journal of Biological Chemistry 107:629-634.

Miller GL (1959) Use of dinitrosalicylic acid reagent for determination of reducing sugar. Analytical Chemistry 31(3):426-428.

Mothes G, Schnorpfeil C, Ackermann JU (2007) Production of PHB from crude glycerol. Engineering in Life Sciences 7(5):475-479.

Moulik SP (1971) Physical aspects of electrofiltration. Environmental Science & Technology 5(9):771-776.

Mukai Y, Yamaguchi S, Kime H, Iritani E (2009) Dead-end ultrafiltration characteristics of particulate suspensions containing macromolecule. Kagaku Kogaku Ronbunshu 35(1):87-93.

Mullon C, Radovich JM, Behnam B (1985) A semiempirical model for electroultrafiltration-diafiltration. Separation Science and Technology 20(1):63-72.

Nakamura H (1996) Roles of electrostatic interaction in proteins. Quarterly Reviews of Biophysics 29(1):1-90.

Nakano T, Nakano K, Sim JS (1994) A simple rapid method to estimate hyaluronic-acid concentrations in rooster comb and wattle using cellulose-acetate electrophoresis. Journal of Agricultural and Food Chemistry 42(12):2766-2768.

Nakashita H, Arai Y, Yoshioka K, Fukui T, Doi Y, Usami R, Horikoshi K, Yamaguchi I (1999) Production of biodegradable polyester by a transgenic tobacco. Bioscience Biotechnology and Biochemistry 63(5):870-874.

Naumann D (2001) Composition and structure of complex biological material, in: Gremlich HU and Yan B (Ed.), Infrared and Raman Spectroscopy of Biological Materials, Particle Spectroscopy Series Vol. 24, Marcel Dekker, New York, pp. 324-377.

Neesse TH, Dueck J, Djatchenko E (2009) Simulation of filter cake porosity in solid/liquid separation. Powder Technology 193(3):332-336.

Nge KL, Nwe N, Chandrkrachang S, Stevens WF (2006) Chitosan as a growth stimulator in orchid tissue culture. Plant Science 170(6):1185-1190.

Noble PW, Lake FR, Henson PM, Riches DWH (1993) Hyaluronate activation of Cd44 induces insulin-like growth factor-I expression by a tumor-necrosis-factor-alpha dependent mechanism in murine macrophages. Journal of Clinical Investigation 91(6):2368-2377.

Ogrodowski CS, Hokka CO, Santana MHA (2005) Production of hyaluronic acid by *Streptococcus*. Applied Biochemistry and Biotechnology 121:753-761.

Ohshima H (2006) Electrokinetic behavior of particles: Theory, in: Somasundan P (Ed.), Encyclopedia of Surface and Colloid Science, Vol.3, CRC Press, Boca Raton pp. 2143-2160.

Ondruschka J, Trutnau M, Bley T (2008) Gewinnung und Potenziale des Biopolymers Chitosan. Chemie Ingenieur Technik 80(6):811-820.

Opong WS, Zydney AL (1991) Hydraulic permeability of protein layers deposited during ultrafiltration. Journal of Colloid and Interface Science 142(1):41-60.

Oregan M, Martini I, Crescenzi F, Deluca C, Lansing M (1994) Molecular mechanisms and genetics of hyaluronan biosynthesis. International Journal of Biological Macromolecules 16(6):283-286.

Orsat V, Raghavan GSV, Sotocinal S, Lightfoot DG, Gopalakrishnan S (1999) Roller press for electro-osmotic dewatering of bio-materials. Drying Technology 17(3):523-538.

Park YG (2005) Improvement of dead-end filtration during crossflow electro-microfiltration of proteins. Journal of Industrial and Engineering Chemistry 11(5):692-699.

Patnaik PR (2005) Perspectives in the modeling and optimization of PHB production by pure and mixed cultures. Critical Reviews in Biotechnology 25(3):153-171.

Pfenning K, Bunge RP (1974) Freeze fracturing of nerve growth cones and young fibers - study of developing plasma membrane. Journal of Cell Biology 63(1):180-196.

Poirier Y, Dennis DE, Klomparens K, Somerville C (1992) Polyhydroxybutyrate, a biodegradable thermoplastic, produced in transgenic plants. Science 256(5056):520-523.

Poirier Y, Nawrath C, Somerville C (1995) Production of polyhydroxyalkanoates, a family of biodegradable plastics and elastomers, in bacteria and plants. Bio-Technology 13(2):142-150.

Rane KD, Hoover DG (1993) Production of chitosan by fungi. Food Biotechnology 7(1):11-33.

Rebah FB, Yan S, Filali-Meknassi Y, Tyagi RD, Suramppalli RY (2004) Bacterial production of bioplastics, in: Surampalli R and Tyagi RD (Ed.) Advances of Water and Wastewater Treatment. American Society of Civil Engineers, USA, pp. 42-71.

Reddy CSK, Ghai R, Kalia VC (2003) Polyhydroxyalkanoates: an overview. Bioresource Technology 87(2):137-146.

Riis V, Mai W (1988) Gas chromatographic determination of polyhydroxybutyric acid in microbial biomass after hydrochloric acid propanolysis. Journal of Chromatography 445(1):285-289.

Rinaudo M (2006) Chitin and chitosan: Properties and applications. Progress in Polymer Science 31(7):603-632.

Russotti G, Göklen KE (2001) Cross-flow Membrane Filtration of Fermentation Broth, in: Wang WK (Ed.), Marcel Dekker, Inc., New York.

Sambrook J, Fritsch EF, Maniatis T (1989) Molecular cloning: A laboratory manual. Cold Spring Harbor Press, Cold Spring Harbor, USA.

Saveyn H, Van der Meeren P, Hofmann R, Stahl W (2005) Modelling two-sided electrofiltration of quartz suspensions: Importance of electrochemical reactions. Chemical Engineering Science 60(23):6768-6779.

Schlegel HG, Kaltwasser H, Gottschalk G (1961) Ein Submersverfahren Zur Kultur Wasserstoffoxydierender Bakterien - Wachstumsphysiologische Untersuchungen. Archiv für Mikrobiologie 38(3):209-222.

Schwuger MJ (1996) Lehrbuch der Grenzfläch. Thieme, New York.

Shaw DJ (1992) Introduction to Colloid and Surface Chemistry. Butterworth-Heinemann Ltd., Oxford.

Shimahara K, Takiguchi Y, Kobayashi T, Uda K, Sannan T (1989) Screening of mucoraceae strains suitable for chitosan production, in:

Skjak-Braek G, Anthonsen T, Sandford P. (Ed.), Chitin and Chitosan, Elsevier, Amsterdam, pp. 171-183.

Slater S, Mitsky TA, Houmiel KL, Hao M, Reiser SE, Taylor NB, Tran M, Valentin HE, Rodriguez DJ, Stone DA, Padgette SR, Kishore G, Gruys KJ (1999) Metabolic engineering of Arabidopsis and Brassica for poly(3-hydroxybutyrate-co-3-hydroxyvalerate) copolymer production. Nature Biotechnology 17(10):1011-1016.

Smith PK, Krohn RI, Hermanson GT, Mallia AK, Gartner FH, Provenzano MD, Fujimoto EK, Goeke NM, Olson BJ, Klenk DC (1985) Measurement of protein using bicinchoninic acid. Analytical Biochemistry 150(1):76-85.

Sousa AS, Guimaraes AP, Goncalves CV, Silva IJ, Cavalcante CL, Azevedo DCS (2009) Purification and characterization of microbial hyaluronic acid by solvent precipitation and size-exclusion chromatography. Separation Science and Technology 44(4):906-923.

Steinbüchel A (2003) Biopolymers. Vol. 1-10, Wiley-VCH, Weinheim.

Steinbüchel A, Doi Y (2005) Biotechnology of Biopolymers. Vol. 1, Wiley-VCH, Weinheim.

Synowiecki J, Al-Khateeb NAAQ (1997) Mycelia of *Mucor rouxii* as a source of chitin and chitosan. Food Chemistry 60(4):605-610.

Tadros T (2007) General principles of colloid stability and the role of surface forces, in: Tadros T (Ed.), Colloid Stability, Wiley-VCH Verlag, Weinheim.

Tamer IM, Moo-Young M, Chisti Y (1998) Disruption of *Alcaligenes latus* for recovery of poly(beta-hydroxybutyric acid): Comparison of high-pressure homogenization, bead milling, and chemically induced lysis. Industrial & Engineering Chemistry Research 37(5):1807-1814.

Tan SC, Tan TK, Wong SM, Khor E (1996) The chitosan yield of *Zygomycetes* at their optimum harvesting time. Carbohydrate Polymers 30(4):239-242.

Tavares LZ, da Silva ES, Pradella JGD (2004) Production of poly(3-hydroxybutyrate) in an airlift bioreactor by *Ralstonia eutropha*. Biochemical Engineering Journal 18(1):21-31.

Teng WL, Khor E, Tan TK, Lim LY, Tan SC (2001) Concurrent production of chitin from shrimp shells and fungi. Carbohydrate Research 332(3):305-316.

Trutnau M, Suckale N, Groeger G, Bley T, Ondruschka J (2009) Enhanced chitosan production and modeling hyphal growth of *Mucor rouxii* interpreting the dependence of chitosan yields on processing and cultivation time. Engineering in Life Sciences 9(6):437-443.

Valentin HE, Broyles DL, Casagrande LA, Colburn SM, Creely WL, DeLaquil P A, Felton HM, Gonzalez KA, Houmiel KL, Lutke K, Mahadeo DA, Mitsky TA, Padgette SR, Reiser SE, Slater S, Stark DM, Stock RT, Stone DA, Taylor NB, Thorne GM, Tran M, Gruys KJ (1999) PHA production, from bacteria to plants. International Journal of Biological Macromolecules 25(1-3):303-306.

Vasconcelos HL, Camargo TP, Goncalves NS, Neves A, Laranjeira MCM, Favere VT (2010) Chitosan crosslinked with a metal complexing agent: Synthesis, characterization and copper(II) ions adsorption. Reactive & Functional Polymers 68(2):572-579.

Verwey EJW, Overbeek JTG (1948) Theory of the stability of lyphobic colloids. Elsevier Amsterdam.

Wang XY, Peng SW, Dong LS (2005) Effect of poly(vinyl acetate) (PVAc) on thermal behavior and mechanical properties of poly(3-hydroxybutyrate)/poly(propylene carbonate) (PHB/PPC) blends. Colloid and Polymer Science 284(2):167-174.

Weber K, Stahl W (2002) Improvement of filtration kinetics by pressure electrofiltration. Separation anc Purification Technology 26(1):69-80.

White SA, Farina PR, Fulton I (1979) Production and isolation of chitosan from *Mucor rouxii*. Applied and Environmental Microbiology 38(2):323-328.

Yang RJ, Li SQ, Zhang QH (2004) Effects of pulsed electric fields on the activity of enzymes in aqueous solution. Journal of Food Science 69(4):C241-C248.

Yang XR, Yuan XY, Cai DN, Wang SY, Zong L (2009) Low molecular weight chitosan in DNA vaccine delivery via mucosa. International Journal of Pharmaceutics 375(1-2):123-132.

Yen MT, Yang JH, Mau JL (2009) Physicochemical characterization of chitin and chitosan from crab shells. Carbohydrate Polymers 75(1):15-21.

Yoo HS, Lee EA, Yoon JJ, Park TG (2005) Hyaluronic acid modified biodegradable scaffolds for cartilage tissue engineering. Biomaterials 26(14):1925-1933.

Yukawa H, Kobayahi K, Tsukui Y, Yamano S, Iwata M (1976) Analysis of batch elektrokinetic filtration. Journal of Chemical Engineering of Japan 9:396-401.

Yukawa H, Shimura K, Suda A, Maniwa A (1983) Cross-flow electro-ultrafiltration for colloidal solutions of proteins. Journal of Chemical Engineering in Japan 16(4):305-311.

Zamani A, Edebo L, Sjostrom B, Taherzadeh MJ (2007) Extraction and precipitation of chitosan from cell wall of *Zygomycetes* fungi by dilute sulfuric acid. Biomacromolecules 8(12):3786-3790.

Zhang K, Helm J, Peschel D, Gruner M, Groth T, Fischer S (2010) NMR and FT–Raman characterisation of regioselectively sulfated chitosan regarding the distribution of sulfate groups and the degree of substitution. Polymer 51(21):4698-4705.

Zhou HD, Ni JR, Huang W, Zhang JD (2006) Separation of hyaluronic acid from fermentation broth by tangential flow microfiltration and ultrafiltration. Separation and Purification Technology 52(1):29-38.

Appendices A – Experimental data

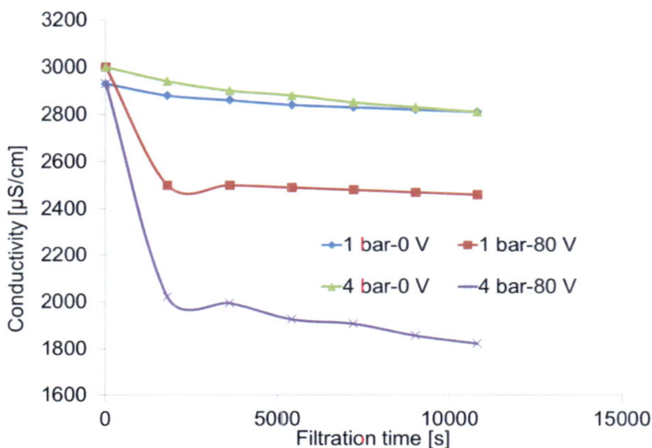

Appendix A1: Conductivity-time diagram for xanthan dispersed without salt addition (~250 µS/cm), electrofiltrated using buffer solution with a conductivity of ~3000 µS/cm, applied pressure of 1 bar, 4 bars and voltage of 0 V, 80 V

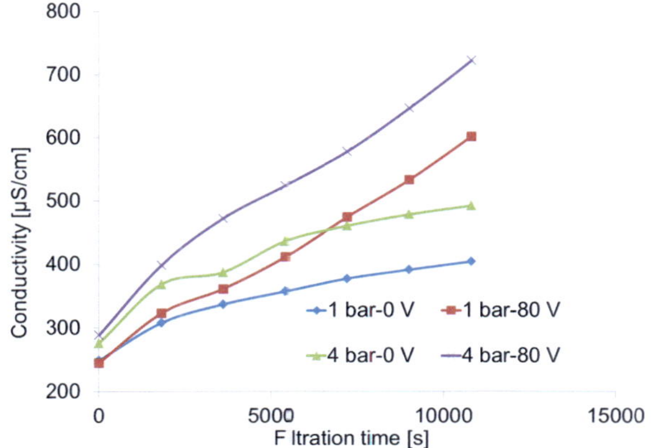

Appendix A2: Conductivity-time diagram for xanthan dispersed with salt addition (~3000 µS/cm), electrofiltrated using buffer solution with a conductivity of ~250 µS/cm, applied pressure of 1 bar, 4 bars and voltage of 0 V, 80 V

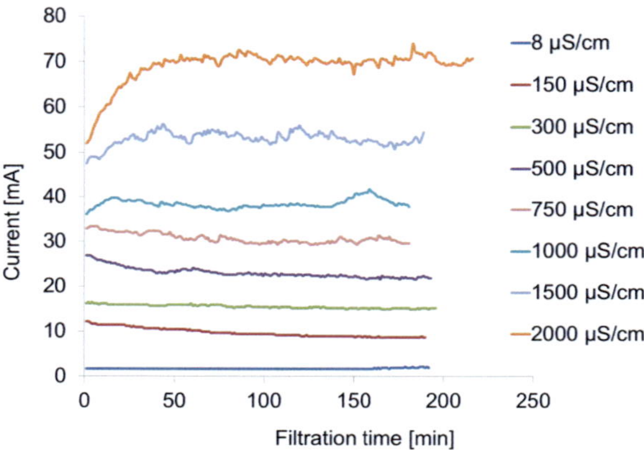

Appendix A3: Current progress of the experiments conducted using different conductivities of buffer solution

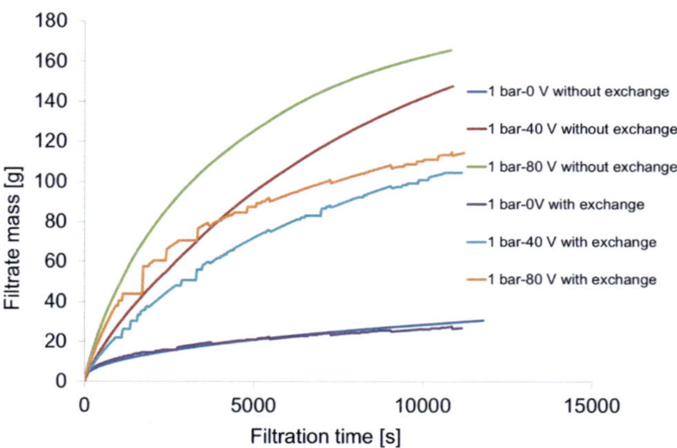

Appendix A4: Filtrate mass-time diagram for the experiments with and without exchange of the filtrate

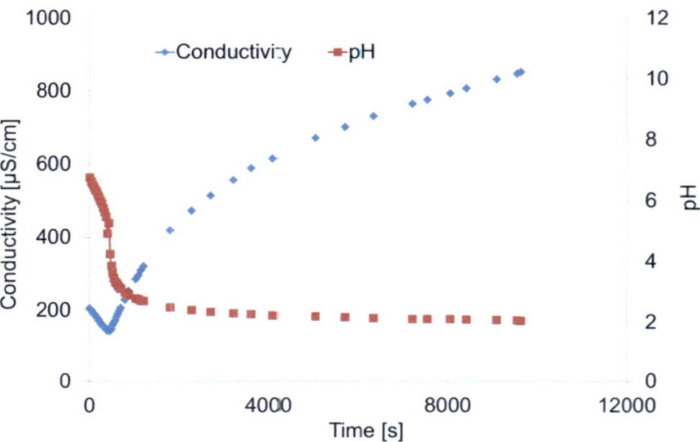

Appendix A5: The change of conductivity and pH in the flushing solution collected on anode side vessel

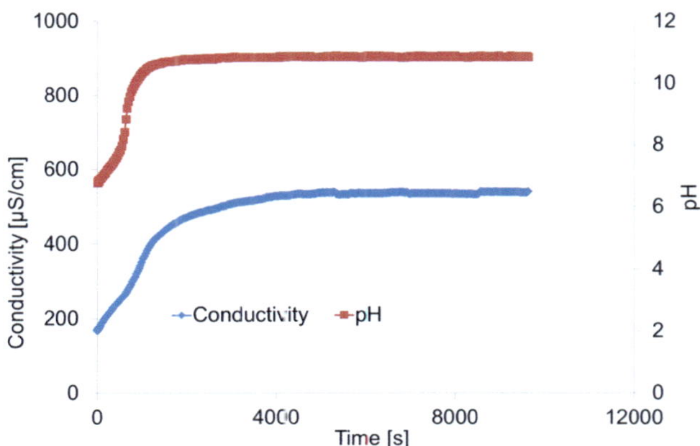

Appendix A6: The change of conductivity and pH in the flushing solution collected on cathode side vessel

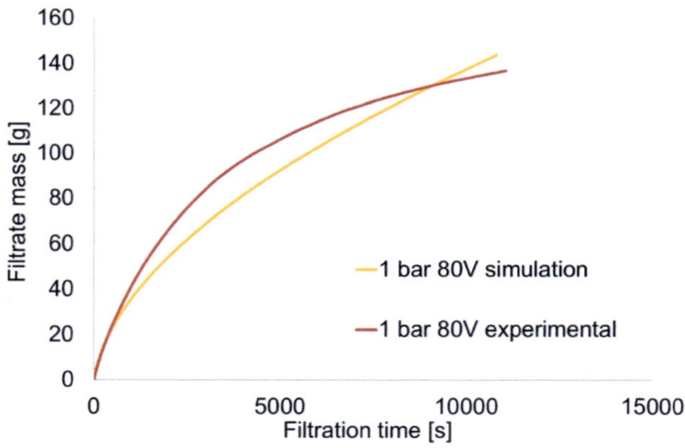

Appendix A7: Comparison between simulated and experimented results at 40 V

Appendices B – Technical Sketches

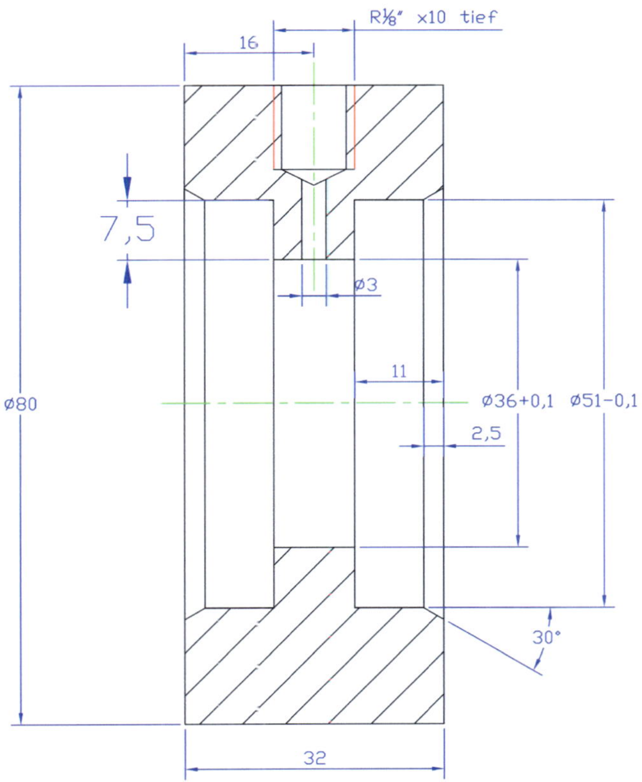

Datum	Name	Maße ohne Tolerarzangabe nach DIN 7:68 m	Bezeichnung		
03.09.08	Steinweg		Filterkammer		

Maßstab	Anzahl	Werkstoff	Bemerkung	Datei	Kostenstelle
2:1	1	Polycarbcnat Makrolon	–	Filterkammer .dwg	5001

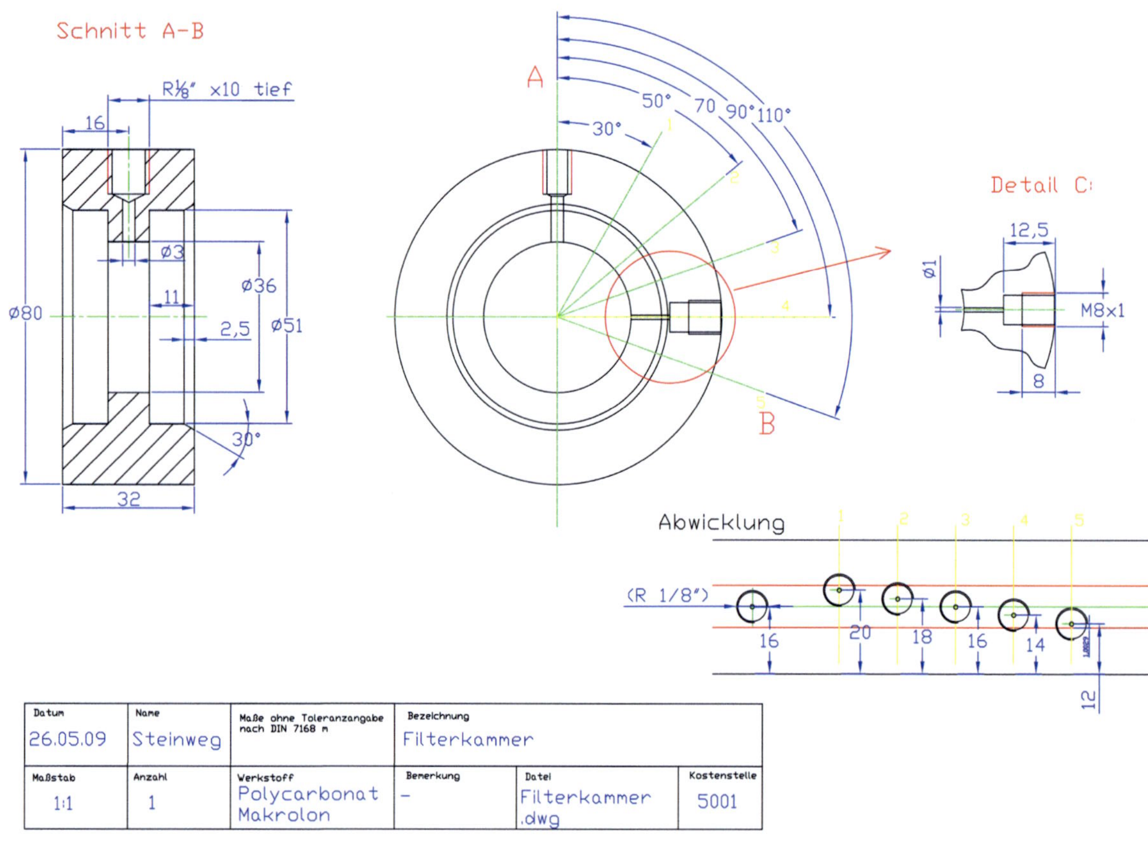

Datum	Name	Maße ohne Toleranzangabe nach DIN 7168 m	Bezeichnung		
26.05.09	Steinweg		Filterkammer		

Maßstab	Anzahl	Werkstoff	Bemerkung	Datei	Kostenstelle
1:1	1	Polycarbonat Makrolon	–	Filterkammer .dwg	5001

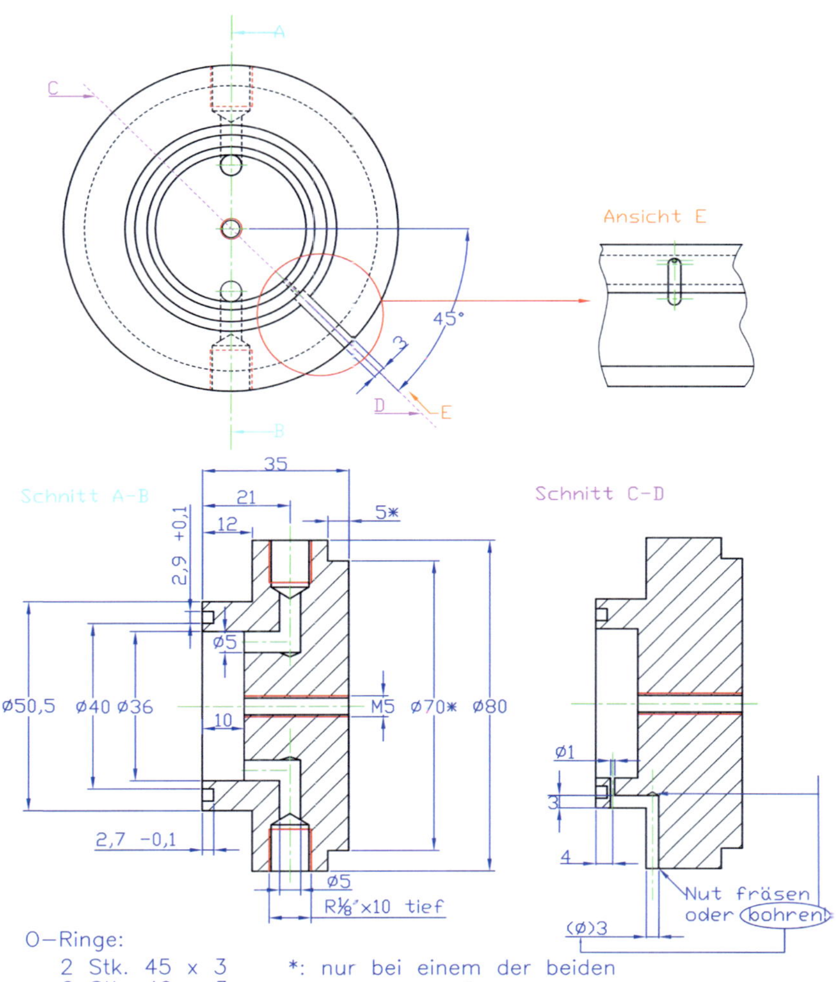

Datum	Name	Maße ohne Toleranzangabe nach DIN 7168 m	Bezeichnung		
09.10.09	Serve		Spülkammern		
Maßstab	Anzahl	Werkstoff	Bemerkung	Datei	Kostenstelle
1:1	2	Plexiglas	–	Filterkammer .dwg	8156

Appendices C – List of symbols and abbreviations

Latin symbols

a	Colloidal particle radius
A	Filtration surface area
A_H	Hamaker constant
c	Molarity
E	Electric field
E_{crit}	Critical electric field strength
f	Concentration factor
F_E	Electrophoretic force
F_W	Hydrodynamic resistance force
G_A	Van der Waals energy of attraction
G_{el}	Electrostatic energy of repulsion
M	Molecular weight
m	Sample mass
$m_{cake,dry}$	Dry mass of filter cake
N	Ionic mobile species
n_i^∞	Bulk concentration
pCO_2	Partial carbon dioxide pressure
pO_2	Partial oxygen pressure
q	Net charge of the particle
s	Compressibility of the cake
T	Absolute temperature
t_f	Filtration time

u	Mobility of particle
U	Voltage drop
V_f	Filtrate volume
z_i	valance

Greek symbols

ε_r	Relative permittivity
α_{av}	Average specific filter cake resistance
v	Velocity of the particle
Λ	Electrolyte conductivity
ΔP	Hydrostatic pressure
Δp_E	Electro-osmotic pressure
$\Delta\psi,$	Potential difference across the membrane
κ	Debye-Hückel parameter
ψ_0	Surface potential
ψ_d	Stern potential
ζ	Zeta potential
η	Dynamic viscosity

Abbreviations

3-HBME	(R)-3-hydroxybutyric acid methyl ester
BCA	Bicinchoninic Acid Assay
BMBF	Federal Ministry of Education and Research
BSA	Bovine Serum Albumin
D_{DA}	Degree of Deacetylation
DSMZ	German Collection of Microorganisms and Cell Cultures

EF	Electrofiltration
FITC	Fluorescein Isothiocyanate
FT	Fourier Transform
GC	Gas Chromatography
GlcUA	Glucuronic acid
GlcNAc	N-acetylglucosamine
GPC	Gel Permeations Chromatography
HA	Hyaluronic acid
ICP	Inductively Coupled Plasma
IR	Infrared
LED	Light Emitting Diode
MALLS	Multi-Angle Laser Light Scattering
MF	Microfiltration
MS	Mass Spectrometry
OD	Optical Density
OES	Optical Emission Spectrometry
PES	Polyethersulfone
PHA	Polyhdroxyalkonate
PHB	Polyhydroxybutyrate
PVDF	Polyvinylidene fluoride
RI	Refractive index
RO	Reverse osmosis
SEC	Size Exclusion Chromatography
SEM	Scanning Electron Microscope
SIAB	Saxon Institute of Applied Biotechnology
UF	Ultrafiltration
UV	Ultraviolet

Constants

e	Elementary electric charge	1.602×10^{-19} C
ε_0	Permittivity of vacuum	8.85×10^{-12} A s V^{-1} m^{-1}
k	Boltzman constant	1.38×10^{-23} J K^{-1}